Deep Learning Unveiled: Mastering Ethical and Explainable AI

Deep Learning Unveiled: Mastering Ethical and Explainable AI

Alexandra Thompson

index

Introduction

1. The Rise of Deep Learning: A Brief Overview

In the modern era of technology, few advances have caught the imagination and impacted businesses as profoundly as deep learning. Often referred to as the crown jewel of artificial intelligence (AI), deep learning has ushered in a new era of possibilities, from self-driving cars to medical diagnosis. This chapter presents a succinct yet thorough account of the astonishing emergence of deep learning, tracing its roots, important milestones, and the transformational impact it continues to have on numerous areas.

Origins of Evolution

The foundation of deep learning may be traced back to the 1940s when the notion of artificial neural networks (ANNs) was first established. Inspired by the structure and function of the human brain, early neural networks tried to imitate biological processes in order to tackle complicated issues. However, the computer resources available at the time constrained the scale and complexity of these networks, leading to minimal advancement.

Fast forward to the 21st century, increases in computer capacity, the availability of big data, and breakthroughs in algorithmic innovation coincided to ignite the revival of deep learning. The advent of deep neural networks, typified by numerous layers of interconnected nodes, represented a turning point. With the emergence of Convolutional Neural Networks (CNNs) and Recurrent Neural Networks (RNNs), the discipline gained interest in several applications, particularly computer vision and natural language processing.

Key Milestones

One of the important occasions that brought deep learning into the forefront was the ImageNet Large Scale Visual Recognition Challenge (ILSVRC) in 2012. The emergence of deep neural networks, particularly Geoffrey Hinton's pioneering AlexNet, greatly increased image categorization accuracy and demonstrated the transformative potential of deep learning. This victory not only inspired future research but also paved the door for large-scale deployment of deep learning in computer vision applications.

Subsequent years witnessed a flurry of advances. In natural language processing, the advent of word embeddings like Word2Vec and the development of Transformer models, including as BERT and GPT, transformed the discipline. These models used the power of deep learning to analyse and generate human-like text, opening opportunities to applications like language translation, sentiment analysis, and even creative writing.

Transformative Impact

The disruptive impact of deep learning is felt throughout industries, from healthcare to banking, entertainment to transportation. In healthcare, deep learning models are boosting diagnostic accuracy in medical imaging, speeding medication discovery, and enabling tailored treatment plans. For instance, in radiology, CNNs have exhibited extraordinary proficiency in spotting defects in X-rays and MRIs, potentially reducing human error and increasing patient outcomes.

The automotive industry is facing a paradigm shift with the development of autonomous vehicles, powered by deep learning algorithms. These vehicles rely on intricate neural networks to evaluate massive volumes of sensor data in real time, making split-second judgements to negotiate difficult traffic circumstances. The consequences for road safety and transportation efficiency are enormous.

Challenges and Future Prospects

However, the rise of deep learning is not without obstacles. The "black-box" aspect of some deep neural networks has prompted questions regarding transparency and interpretability. While deep learning models can attain incredible accuracy, understanding the rationale behind their conclusions remains a difficulty. This constraint has generated a parallel field of research: explainable AI. Researchers are striving to develop ways that shed light on the inner workings of these networks, revealing insights into how decisions are made.

Furthermore, the ethical consequences of deep learning's increasing incorporation into society cannot be disregarded. Bias in training data can lead to biased models, prolonging unfair outcomes and reinforcing societal injustices. Ensuring that deep learning models are trained on varied and representative datasets is a vital step in reducing these biases.

Conclusion

In conclusion, the growth of deep learning has been a journey distinguished by tenacity, ingenuity, and transformational potential. From its humble origins in artificial neural networks to its current stature as the backbone of modern AI, deep learning has changed how we approach tough issues and unearth unexpected insights. As we move forward, the concerns of transparency, ethics, and bias must be faced hand in hand with technology breakthroughs. The deep learning revolution has set the stage for a future where the bounds of what is conceivable continue to expand, altering industries and enhancing our lives in ways formerly thought to be the province of science fiction.

2. The Need for Ethical and Explainable AI

In the field of artificial intelligence (AI), where machines are supposed to emulate human cognitive functions, ethical problems have taken center stage. The integration of AI into various facets of our life has brought about a vital need for ensuring that these technologies align with our values, respect human rights, and contribute constructively to society. As AI gets increasingly complicated and autonomous, the desire for ethical and explainable AI grows louder. This chapter looks into the significance of ethical and explainable AI, addressing the rationale behind these imperatives and the issues they strive to overcome.

The Promise and Peril of AI

The fast breakthroughs in AI have given rise to great promise and potential. AI systems are powering medical diagnosis, streamlining supply chains, transforming customer service, and even assisting in scientific discoveries. However, this transforming potential comes with a double-edged sword. While AI might boost efficiency, accuracy, and convenience, it can also perpetuate biases, infringe upon privacy, and pose unforeseen threats.

Consider the instance of a self-driving car that must make rapid decisions in potentially life-threatening situations. How should the car's algorithm prioritize the safety of the occupants vs pedestrians? Should it benefit the young over the elderly? These ethical dilemmas emphasise the challenge of incorporating human values into AI systems.

The Ethical Imperative

The ethical imperative in AI is around ensuring that AI systems align with human values, ideals, and society norms. Without sufficient management, AI programmes can unwittingly discriminate against some groups, intensify existing prejudices, and jeopardise fairness. A typical example is the instance of AI-based hiring tools that demonstrated gender prejudice due to biased training data, leading to the reinforcement of gender gaps in the workplace.

To address these difficulties, ethical frameworks for AI development are being established. These frameworks emphasize concepts such as fairness, transparency, accountability, and human-centered design. Implementing these principles needs thorough testing, continual monitoring, and a dedication to resolving unexpected consequences.

The Quest for Explainability

Explainable AI (XAI) is a closely related idea that addresses the requirement for transparency and explanation in AI decision-making. Many AI models, particularly deep neural networks, operate as "black boxes," meaning their decision-making processes are not immediately intelligible by humans. This lack of openness can be problematic, especially in high-stakes applications where accountability is critical.

Imagine a case when an AI rejects a person a loan or suggests a medical treatment, but cannot provide a coherent rationale for its choice. This opacity reduces trust in AI and can lead to biased or unfair outcomes. Explainable AI intends to overcome this gap by enabling humans to comprehend the rationale behind AI decisions, promoting confidence and enabling prompt interventions when necessary.

The Challenges of Ethical and Explainable AI

Despite the compelling need for ethical and explainable AI, accomplishing these aims is not without hurdles. One big challenge is establishing the correct balance between innovation and regulation. Overregulation can hamper technological advancement, while underregulation can lead to ethical violations. Finding the medium ground involves collaboration among governments, industries, researchers, and ethicists.

Another problem is creating universal ethical principles that apply across cultures and settings. What is considered ethical in one culture could not align with another's values. Navigating this cultural diversity takes sensitivity and a global perspective.

Toward a Responsible AI Future

While the hurdles are enormous, the pursuit of ethical and explainable AI is a moral and practical imperative. As AI systems become more integrated with our lives, they influence our actions, change our perceptions, and impact our futures. The obligation to guarantee that these technologies follow ethical norms and provide transparency is on developers, policymakers, and society at large.

Ethical and explainable AI are not restrictions; they are enablers of a more inclusive, just, and trustworthy AI ecosystem. They empower us to leverage the potential of AI while reducing its perils. As we continue to push the boundaries of AI innovation, we must do so with a profound commitment to the values that define us as human beings.

Conclusion

In conclusion, the rise of AI confronts us with both unparalleled potential and formidable problems. Ethical considerations underline the need of integrating AI systems with human values, eliminating biases, and encouraging fairness. Explainable AI, in turn, strengthens our ability to comprehend and trust AI judgements, providing openness and accountability. As we traverse this complex world, the need for ethical and explainable AI is paramount. The choices we make now will influence the direction of AI's impact on our society, economy, and well-being for years to come.

3. Setting the Stage: Introducing the Core Themes

In the ever-evolving environment of technology, few fields have caught the imagination and impacted industry as profoundly as artificial intelligence (AI). Amidst the remarkable breakthroughs in AI, one particular subgroup stands out: deep learning. As this extraordinary subject continues to transform how we engage with technology, comprehend data, and solve complicated problems, it is vital to develop a strong foundation by introducing its basic ideas. This chapter serves as a guide, presenting the core concepts that drive deep learning and giving a path for the exploration that lies ahead.

Defining Deep Learning

Deep learning, at its heart, is a subfield of AI that tries to emulate the human brain's neural networks in order to analyse and understand data. Its defining feature is the utilisation of deep neural networks, which are made of interconnected layers of nodes, or artificial neurons. These networks can independently learn to complete tasks by altering the strengths of connections between nodes, a process known as training.

Deep learning models have showed great proficiency in activities that traditionally required human knowledge, from picture identification and language translation to playing sophisticated games. The natural power of deep neural networks to discover intricate patterns across enormous datasets has changed several industries, promising new heights of efficiency, accuracy, and innovation.

The Data-Driven Paradigm

Central to the success of deep learning is its data-driven nature. Traditional rule-based systems relied significantly on human-crafted algorithms, restricting their flexibility to new conditions. Deep learning, however, thrives on enormous volumes of data, allowing models to learn from examples rather than explicit instructions. This data-centric strategy helps AI to generalize from experience, making it capable of handling previously unencountered difficulties.

Consider the instance of natural language processing, where deep learning models like Transformers are trained on enormous text datasets. These models can generate coherent and contextually relevant text, revealing a peek of AI's capacity to interpret human language in ways that were formerly believed implausible.

The Power of Neural Networks

At the heart of deep learning lies the neural network, a computer design inspired by the linked neurons in the human brain. Neural networks comprise of layers: an input layer takes data, hidden layers process it, and an output layer generates the desired result. These layers comprise nodes, each of which performs a mathematical action on incoming data.

Hidden inside the layers of these networks are weights, or parameters, that affect the strength of connections between nodes. During training, these weights are modified by a process called backpropagation, where the model learns to minimize the discrepancy between its predictions and the actual results. This iterative learning process, powered by large volumes of data, allows deep learning models to fine-tune their internal representations and produce increasingly accurate predictions.

Challenges on the Horizon

Despite its tremendous successes, deep learning is not without obstacles. One key challenge is the necessity for enormous amounts of tagged data. Supervised learning, the dominant paradigm in deep learning, depends on labeled samples to train models. Acquiring and annotating data for multiple domains can be time-consuming, costly, and even infeasible un certain situations.

Another challenge is the tendency of deep neural networks to become "black boxes." The complicated linkages and transformations occurring inside these networks can be difficult to grasp, leading to a lack of transparency in decision-making. This opacity raises questions about accountability, prejudice, and ethical considerations, especially in high-stakes applications like healthcare and finance.

The Path Forward

As we continue on this investigation of deep learning's key themes, it's vital to keep in mind the dynamic nature of the discipline. New designs, methods, and approaches are always evolving, pushing the frontiers of what is achievable. It's a sphere where invention is unceasing, fuelled by the joint efforts of researchers, engineers, and pioneers.

This chapter lays the foundation for a deeper look into the intricacies of deep learning, from its ethical implications and explainability concerns to its role in altering industries and addressing complicated problems. As we go through the complexity of this transformative discipline, we'll discover the layers that create its foundation, illuminating the mechanisms that power its extraordinary powers and examining the routes that lead to a more educated, ethical, and innovative future.

Chapter 1

Foundations of Deep Learning

1.1 Understanding Neural Networks and Their Components

In the field of artificial intelligence, neural networks have emerged as the workhorses of current machine learning. Inspired by the intricate interconnections of neurons in the human brain, neural networks are at the heart of various AI applications, driving advancements in image identification, natural language processing, and much more. This chapter digs into the core notion of neural networks, demystifying their components and operations, and shining light on how these computational constructions have changed the field of AI.

The Neural Network Paradigm

At its core, a neural network is a computational model designed to process data and generate predictions by imitating the action of interconnected neurons. This concept draws inspiration from the biological brain, where neurons connect through electrical signals, producing intricate networks that underlie human cognition. The goal of artificial neural networks is to mimic this dense web of connections in a mathematical framework.

A neural network has layers of interconnected nodes, or artificial neurons. Each neuron receives information, processes it using a mathematical function, and creates an output. The strength of connections between neurons, known as weights, is what allows the network to understand patterns and relationships within the data.

Layers and Architecture

Neural networks are arranged into layers, each providing a specific purpose in the data processing pipeline. The three major types of layers are the input layer, hidden layers, and the output layer. The input layer receives raw data, which is then processed and modified as it moves through the hidden levels. The output layer delivers the final prediction or categorization.

The architecture of a neural network relates to the layout and number of layers and neurons inside those layers. Deep neural networks, which include numerous hidden layers, have gained significance due to their capacity to capture nuanced patterns in data. Shallow networks, on the other hand, have fewer hidden layers and are commonly utilised for simpler tasks.

Activation Functions

Activation functions play a critical part in defining the output of a neuron. They introduce non-linearity to the network, helping it to grasp complicated correlations within the data. Common activation functions include the sigmoid, ReLU (Rectified Linear Unit), and tanh (hyperbolic tangent) functions. Each activation function has its own properties, impacting how the neuron responds to input data and contributes to the broader network's activity.

Activation functions are crucial in deep learning, enabling networks to represent sophisticated operations and generate accurate predictions. The choice of activation function can effect how quickly the network learns and if it can overcome common training challenges like disappearing gradients.

Weight Adjustments and Learning

The crux of neural network learning resides in altering the weights that connect neurons inside the network. This modification is performed by a technique called as backpropagation. Backpropagation includes comparing the network's anticipated output with the actual target output, determining the error, and propagating this error backward through the layers. The weights are then modified in a way that decreases the error, matching the network's predictions with the desired outcomes.

This repeated process of forward and backward sweeps, coupled with weight modifications, allows the neural network to update its internal representations and improve its predictions over time. The complexity of deep neural networks, with their various layers and weights, enables them to learn subtle patterns from enormous datasets.

Training and Optimization

Training a neural network includes exposing it to a wide set of instances and allowing it to alter its weights through backpropagation. The procedure needs careful consideration of hyperparameters, such as learning rate and batch size, which impact how quickly the network converges to optimal weights.

Optimization techniques, such as stochastic gradient descent and its derivatives, guide the weight changes during training. These strategies ensure that the network doesn't get stuck in local minima and converges to a solution that minimizes the prediction error.

Conclusion

In conclusion, knowing neural networks and their components is crucial to grasping the mechanisms that fuel modern AI. These computational frameworks, inspired by the complexity of the human brain, have enabled significant progress in activities that were formerly thought hard for robots. From image recognition to language translation, neural networks continue to push the frontiers of what AI can achieve. As we delve deeper into their subtleties, we reveal a world of interconnected nodes, mathematical functions, and weight modifications that collectively drive the creation of artificial intelligence.

1.2 Training and Optimization Techniques: Navigating the Neural Network Landscape

The path of a neural network from a blank slate to a sophisticated problem solver is a complex and iterative process. At the heart of this trip lies the training phase, where the network learns to generate accurate predictions and discover patterns within data. However, this learning process is not without hurdles. Enter optimization procedures, a set of strategies that direct the network's changes and ensure it reaches its peak performance. This chapter digs into the domain of training and optimization approaches, giving light on their significance, methodology, and the crucial role they play in harnessing the potential of neural networks.

The Training Landscape

Training a neural network entails exposing it to a dataset with known inputs and outputs, allowing it to understand the underlying patterns that connect the two. This method is analogous to educating a model to associate certain inputs with corresponding intended outputs. In other words, the network seeks to minimize the discrepancy between its predictions and the actual events.

The training process is guided by the notion of optimization: modifying the network's internal parameters, known as weights, to obtain the greatest feasible performance on the given task. The accuracy of the network's predictions continuously increases across subsequent iterations, or epochs, as it refines its internal representations of the data.

Gradient Descent: The Foundation of Optimization

At the core of many optimization approaches is the concept of gradient descent. This core principle includes iteratively modifying the network's weights in the direction that lowers the error between predictions and actual outcomes. The gradient, simply a vector of partial derivatives, represents the direction of sharpest growth of a function. In the context of neural networks, it directs towards the direction in which the error is minimal.

Gradient descent techniques come in numerous flavors, including batch, mini-batch, and stochastic gradient descent. In batch gradient descent, the full training dataset is used to compute the gradient in each iteration. Mini-batch gradient descent strikes a compromise by partitioning the dataset into smaller subsets, or mini-batches, which are employed for gradient computation. Stochastic gradient descent takes this further, employing a single training example to calculate the gradient at a time. Each option has its merits and cons in terms of convergence speed, memory needs, and computing efficiency.

Challenges and Variants

While gradient descent provides the backbone of optimization, it's not without obstacles. One key challenge is the potential for the algorithm to get caught in local minima, where further modifications don't lead to improvements in the error. To solve this, optimization approaches have emerged to include momentum, learning rate annealing, and adaptive learning rates.

Momentum introduces a "velocity" term that helps the algorithm overcome flat sections and navigate more effectively toward the global minimum. Learning rate annealing includes decreasing the learning rate over time, allowing the algorithm to start with greater steps and eventually refine its trajectory. Adaptive learning rate algorithms, such as AdaGrad, RMSProp, and Adam, dynamically alter the learning rate for each weight based on their historical gradients.

Regularization: Preventing Overfitting

As neural networks expand in complexity, they run the risk of overfitting: learning to memorize the training data instead of generalizing to new, unseen samples. Regularization techniques come to the rescue by preventing overfitting and boosting the network's ability to generalize.

One popular method of regularization is L2 regularization, often known as weight decay. It adds a penalty term to the loss function that discourages big weights, hence diminishing their impact on the network's predictions. Another strategy is dropout, where randomly selected neurons are "dropped out" during training, effectively lowering the network's reliance on specific neurons and preventing co-adaptation.

Transfer Learning: Leveraging Pretrained Models

In recent years, transfer learning has gained importance as an optimization strategy that accelerates the training process and enhances model performance. Transfer learning involves using a pretrained model as a starting point and fine-tuning it on a specific task. This strategy harnesses the knowledge the pretrained model has gathered from a big dataset, saving time and resources during training.

For instance, a convolutional neural network (CNN) pretrained on a vast image dataset can be fine-tuned for a specific image recognition task. By starting with learnt features and modifying them to suit the job at hand, transfer learning helps networks achieve good performance even with limited data.

Conclusion

In the area of neural networks, training and optimization approaches provide the backbone of success. The journey from start to optimal performance is driven by algorithms like gradient descent, which enable the network to learn patterns and generate accurate predictions. Regularization approaches minimise overfitting, whereas transfer learning employs pretrained models to expedite learning and boost performance. As we continue to navigate the landscape of neural networks, understanding these strategies is vital for unleashing the full potential of these incredible computational creations.

1.3 Deep Learning Architectures: CNNs, RNNs, and Beyond

The rise of artificial intelligence (AI) and its astonishing applications across varied sectors have been substantially fueled by the rapid advancement of deep learning architectures. Among the most influential of these architectures are Convolutional Neural Networks (CNNs) and Recurrent Neural Networks (RNNs). These core structures, each adapted to specific types of data and activities, have transformed the landscape of AI by enabling advancements in image recognition, natural language processing, and beyond. This chapter goes into the complexity of CNNs and RNNs, studying their inner workings, applications, and the exciting realm of developing architectures.

Convolutional Neural Networks (CNNs)

Convolutional Neural Networks, also referred to as CNNs or ConvNets, have transformed the field of computer vision. Designed to interpret and evaluate visual data, CNNs are particularly adept at tasks such as picture categorization, object detection, and image segmentation.

The underlying principle of CNNs resides in their capacity to capture local patterns within a picture. This is performed by convolutional layers, which employ filters or kernels to extract features from tiny parts of the input data. These features, representing diverse visual components like edges, textures, and forms, are then blended in succeeding layers to build higher-level representations.

Pooling layers, another fundamental component of CNNs, minimise the spatial dimensions of the data while keeping essential properties. This method helps the network focus on vital information and minimises the computational complexity of succeeding levels. The combination of convolutional and pooling layers allows CNNs to learn hierarchical representations of visual data, enabling them to recognize complicated patterns with surprising precision.

Applications of CNNs

CNNs have found their stride in different real-world applications. In image classification, models like AlexNet, VGG, and ResNet have achieved unparalleled accuracy in detecting objects within images. This power has pervaded industries like healthcare, where CNNs aid in medical picture analysis, letting radiologists spot anomalies and disorders.

Object identification, another sector where CNNs thrive, entails not only detecting objects inside images but also precisely localizing them. This application has significant implications for driverless vehicles, robots, and surveillance systems.

Furthermore, CNNs have a role in image generation and style transfer, making aesthetically appealing art and changing images to embrace the aesthetic aspects of famous artworks. This cross-pollination between art and technology highlights the versatility and creative possibilities of CNNs.

Recurrent Neural Networks (RNNs)

While CNNs excel in visual data processing, Recurrent Neural Networks (RNNs) specialize in sequential data and time-series analysis. This makes them highly suited for tasks like as natural language processing, speech recognition, and language production.

RNNs handle the difficulty of modeling sequences by inserting feedback loops within the network. This input helps the network to maintain an internal state that collects context from prior time steps, making them capable of understanding patterns and dependencies within sequences.

However, classic RNNs have problems in capturing long-term dependencies because to the "vanishing gradient" problem. This difficulty emerges when the gradients used for weight modifications diminish as they travel backward over time, preventing the network from successfully learning distant links within sequences.

Long Short-Term Memory (LSTM) and Gated Recurrent Units (GRUs)

To alleviate the vanishing gradient problem and boost the capacity of RNNs to capture long-range relationships, architectures like Long Short-Term Memory (LSTM) and Gated Recurrent Units (GRUs) were created.

LSTM networks, comprising of memory cells, gates, and input/output mechanisms, can store and retrieve information over lengthy time periods. This memory retention makes them well-suited for jobs like machine translation, speech recognition, and sentiment analysis, where comprehending the context of the full sequence is crucial.

GRUs, a simpler variant of LSTMs, also address the vanishing gradient issue. They utilize gating methods to restrict the flow of input within the network, allowing them to capture temporal relationships without suffering from the same shortcomings as regular RNNs.

Applications of RNNs

RNNs' capacity to model sequences has led to breakthrough applications. In natural language processing, they excel in tasks such as machine translation, text production, and sentiment analysis. By understanding the context and links between words in a sentence, RNNs may generate coherent text or accurately translate across languages.

RNNs also play a significant role in voice recognition, enabling systems to transcribe spoken words into text. Voice assistants, dictation software, and voice commands in smart devices all rely on the capabilities of RNN-based models to accurately translate audio signals into intelligible text.

Beyond CNNs and RNNs: Emerging Architectures

As the science of deep learning evolves, new architectures are continually developing, each adapted to unique objectives and challenges. Attention mechanisms, a revolutionary notion, have received attention due to their capacity to focus on relevant bits of incoming data. Transformers, a paradigm built on attention mechanisms, have transformed natural language processing, enabling exceptional results in tasks like language translation, language interpretation, and question answering.

Generative Adversarial Networks (GANs), another breakthrough architecture, have revolutionised the area of generative modeling. GANs consist of two neural networks—one that creates data and another that discriminates between actual and produced data. This adversarial training leads to the synthesis of highly realistic synthetic data, with applications ranging from art generation to data augmentation.

Conclusion

The environment of deep learning architectures is a dynamic domain where innovation emerges at an unprecedented pace. Convolutional Neural Networks revolutionized computer vision, whereas Recurrent Neural Networks transformed sequential data analysis. Emerging architectures like Transformers and GANs continue to push the frontiers of what is attainable in AI. As we traverse the intricacies of these systems, we see the potential to transform industries, grasp human perception, and pave the path for a future where robots perceive, interpret, and create in ways formerly deemed the domain of human intellect.

1.4 The Power and Pitfalls of Pretrained Models

In the ever-evolving environment of artificial intelligence, pretrained models have emerged as game-changers, giving a surprising shortcut to obtaining state-of-the-art performance in diverse tasks. These models, pre-trained on enormous datasets and fine-tuned for specific applications, harness the information collected from vast volumes of data to give powerful initializations for subsequent tasks. While pretrained models offer exceptional advantages, they also come with significant hazards that necessitate careful attention. This chapter goes into the intricacies of pretrained models, studying their possibilities, problems, and the delicate balance required to harness their potential successfully.

The Power of Transfer Learning

At the basis of pretrained models lies the notion of transfer learning, which enables knowledge obtained from one task to be transferred and used to another similar task. Transfer learning utilises the idea that the knowledge and patterns learnt from one dataset are likely to have relevance and utility in a different yet related setting.

Pretrained models are frequently trained on enormous datasets for tasks like image classification, natural language processing, or even playing games. These models learn to extract high-level properties, such as edges, forms, and textures in the case of images, or semantic meanings and contextual relationships in text. When fine-tuned on specific tasks with limited datasets, pretrained models display the ability to generalize and adapt their expertise to new problems, often surpassing models trained from scratch.

Advantages of Pretrained Models

The advantages of pretrained models are manifold. First and foremost, they save tremendous time and resources. Instead of building a complex model from the ground up, practitioners can exploit the initializations and feature representations encoded in pretrained models. This is especially beneficial in cases when huge labeled datasets or extensive computational resources are not readily available.

Secondly, pretrained models display extraordinary performance. Their initializations encode a plethora of knowledge from varied datasets, helping them to grasp nuanced patterns and relationships that are tough to learn from little data. This capacity leads to faster convergence and improved accuracy in applications like as picture classification, object detection, and sentiment analysis.

Furthermore, pretrained models increase reproducibility and information exchange. Researchers can openly distribute pretrained models, allowing others to build upon their work and advance the field more rapidly. This exchange of pretrained weights stimulates teamwork and speeds the pace of invention.

Challenges and Considerations

However, the route to properly exploiting pretrained models is not without hurdles. One significant difficulty is domain adaptability. Pretrained models are most effective when the source domain (pretraining data) and the target domain (application data) are closely connected. Significant discrepancies in data distributions might lead to a phenomena called "negative transfer," where the knowledge from the source domain inhibits performance in the target area.

Another consideration is overfitting. While pretrained models provide a head start, fine-tuning them on smaller datasets increases the danger of overfitting. It's vital to find the correct balance between conveying knowledge and adjusting to the specific peculiarities of the task at hand.

Ethical and Bias Concerns

Pretrained models also entail ethical implications, notably in terms of bias propagation. If the original data utilised for pretraining has biases, those biases can be perpetuated in the fine-tuned models. For instance, if a pretrained model is biased toward a certain gender or demographic due to imbalances in the source data, it may mistakenly produce biased results in new circumstances. Addressing and reducing these biases is vital to guarantee that pretrained models contribute to justice and inclusivity.

The Need for Interpretability

A hazard that transcends technical challenges is the lack of interpretability. Pretrained models, especially deep neural networks, are typically dubbed "black boxes" due to their intricate internal representations and transformations. As these models are integrated into vital applications like healthcare and finance, the inability to understand and explain their judgements becomes a big worry. Research in explainable AI is crucial to ensure that pretrained models are transparent and accountable in their predictions.

Conclusion

The value of pretrained models is undeniable, altering the landscape of AI by giving a shortcut to achieving remarkable performance in a wide array of activities. Their capacity to transfer information and generalize patterns from disparate datasets has opened doors to new possibilities and expedited innovation. However, practitioners must overcome the hurdles and risks that come with pretrained models. Ensuring domain adaptability, minimising biases, preventing overfitting, and addressing interpretability concerns are crucial for harnessing the promise of pretrained models effectively and ethically. As the AI field continues to advance, striking the delicate balance between utilising the power of pretrained models and avoiding their flaws will be vital for crafting a responsible and influential AI future.

Chapter 2
The Ethics Imperative

2.1 Ethical Considerations in AI Development

As artificial intelligence (AI) continues to change businesses and reshape the way people engage with technology, ethical questions have taken center stage. The rapid growth of AI brings with it a plethora of benefits and difficulties, ranging from boosting efficiency and convenience to potential biases and societal repercussions. In this chapter, we delve into the vital role of ethical considerations in AI development, addressing the fundamental issues, guiding principles, and the urgent need for responsible AI that fits with human values and upholds the greater good.

The Complexity of Ethical Considerations

Ethical considerations in AI development are complicated and encompass a wide range of fields. At the heart of these issues lay problems about justice, transparency, accountability, prejudice, privacy, safety, and the general influence of AI on society. Addressing these challenges needs a multidisciplinary approach, comprising not only computer scientists and engineers but also ethicists, legislators, social scientists, and representatives from other stakeholder groups.

As AI systems grow more integrated into our daily lives, from tailored suggestions to medical diagnosis, the repercussions of their decisions become increasingly serious. Ensuring that these judgements are made ethically is not merely a matter of compliance but a basic obligation to preserve human rights, societal values, and the well-being of individuals and communities.

Fairness and Bias

One of the major ethical considerations in AI is fairness. AI algorithms are trained on data, and if that data contains prejudices, the resulting AI systems can perpetuate and exacerbate those biases. This can lead to discriminatory outcomes that disproportionately affect particular groups, aggravating existing disparities.

For example, biased training data in predictive police algorithms could lead to the over-policing of underprivileged communities. Similarly, biased recruiting algorithms could perpetuate gender or racial gaps in the workplace. To overcome these challenges, developers must actively try to discover and minimise biases in both training data and algorithms.

Transparency and Accountability

Transparency and accountability are crucial elements of ethical AI development. Users and stakeholders should have a clear knowledge of how AI systems make decisions, what data is used, and how those decisions influence individuals. This transparency creates confidence and encourages people to debate and fix inaccuracies or prejudices.

Furthermore, AI developers and organizations should be liable for the outcomes of their systems. When an AI system makes a mistake or generates a damaging effect, there should be systems in place to repair the issue and prevent similar occurrences in the future. This accountability applies to both technological issues and ethical considerations, ensuring that AI systems are compatible with social norms.

Privacy and Data Security

AI systems frequently rely on enormous volumes of data to make correct predictions and judgements. However, this dependence on data presents serious privacy and data security problems. Collecting, retaining, and processing personal data without informed consent can infringe upon individuals' privacy rights.

To overcome these concerns, AI developers must adhere to severe data protection standards and implement privacy-preserving strategies. This comprises data anonymization, encryption, and data minimization, which ensure that AI systems can work efficiently without compromising individuals' privacy.

Societal Impact and Responsibility

Ethical AI development also demands examining the broader societal impact of AI technologies. Developers have a responsibility to anticipate and mitigate potential harmful outcomes of their creations. For instance, the deployment of driverless vehicles may transform transportation but could potentially harm jobs in the transportation sector. Ethical issues extend to reducing future job displacement through reskilling and workforce development activities.

Additionally, ethical considerations surround the potential misuse of AI technologies. Ensuring that AI systems are not utilised for malevolent reasons, such as deepfake production or monitoring without consent, requires both technical safeguards and effective rules.

Guiding Principles for Ethical AI Development

To traverse the complicated world of ethical AI research, numerous guiding concepts have emerged:

1. Human-Centered Design: AI systems should be developed to boost human capabilities, increase well-being, and respect human autonomy. User demands and values should be at the forefront of development.

2. Fairness and Avoidance of Bias: Developers should aim to minimise biases in data and algorithms to ensure that AI systems treat all individuals fairly and without discrimination.

3. Transparency: AI systems should be clear in their decision-making processes, enabling users to understand how judgements are reached and promoting confidence.

4. Accountability: Developers and organizations should be accountable for the outcomes of their AI systems and should provide methods to fix faults and errors.

5. Privacy and Data Security: AI systems should be built to protect individuals' privacy and data security, ensuring that data is acquired and handled ethically and responsibly.

6. Collaboration and Multidisciplinary Approach: Ethical AI development involves collaboration across many stakeholders, including technologists, ethicists, politicians, and society at large.

Conclusion

In conclusion, ethical considerations in AI development are of fundamental importance as AI technologies continue to transform our environment. The potential of AI brings with it a deep duty to ensure that these technologies accord with human values, protect rights, and contribute positively to society. As we traverse the complexity of AI ethics, it's vital to promote justice, openness, accountability, and privacy in every phase of AI development. By adhering to these principles and engaging in ethical contemplation, we may harness the promise of AI to build a future that is both technologically advanced and morally sound.

2.2 Bias and Fairness Issues in Deep Learning

The rapid growth of deep learning has ushered in a new era of artificial intelligence, with achievements across numerous areas. However, this development is not without its obstacles, and one of the most crucial concerns that has risen to the forefront is bias and fairness in deep learning systems. As AI technologies become increasingly incorporated into our daily lives, it's vital to address the ethical problems regarding biases that can infiltrate these systems. This chapter analyses the complex landscape of bias and fairness issues in deep learning, diving into their origins, repercussions, and the solutions required to construct equitable AI systems that respect diversity and support social norms.

Understanding Bias in Deep Learning

Bias in deep learning refers to the occurrence of systematic and unfair mistakes in AI systems' predictions or conclusions. These mistakes can come from biases existing in the training data, the algorithms themselves, or the judgements made throughout the development process. Biases can emerge in numerous forms, including gender, racial, ethnic, socioeconomic, and cultural biases, among others.

The sources of bias in deep learning are diverse. Training data typically reflects societal biases and inequalities prevalent in the actual world. If the data used for training is distorted or contains preconceptions, the resulting AI model can accidentally perpetuate those biases. Furthermore, biases can be introduced by algorithm design, feature selection, and even labeling judgements during data annotation.

Implications of Bias in AI

The ramifications of bias in AI are far-reaching and can have a dramatic impact on individuals, groups, and society as a whole. Biased AI systems can result in biassed outcomes, reinforcing existing inequities and worsening socioeconomic inequality. For instance, an AI-powered hiring tool that is biased against particular demographic groups can perpetuate discrimination in employment practices.

Bias can also lead to wrong decisions and faulty predictions, affecting the dependability and efficacy of AI systems. Misdiagnoses in medical imaging, biased credit scoring, and misclassification in criminal justice applications are just a few examples of how bias can have major implications.

Moreover, biased AI systems can erode trust and spread unfavourable stereotypes, leading to lower user confidence and skepticism about the fairness of AI technologies. This can limit the general adoption of AI systems and undercut the potential benefits they offer.

Fairness: A Complex Challenge

Achieving fairness in deep learning systems is a challenging challenge that incorporates both technological and ethical factors. Fairness does not have a one-size-fits-all definition; it varies based on context, society standards, and individual perspectives. However, there are some essential concepts that guide the goal of fairness in AI:

1. Equal Opportunity: Fair AI systems should provide equal opportunities and outcomes to all persons, regardless of their history, characteristics, or demographic factors.

2. Group Fairness: AI systems should avoid disproportionately favoring or disadvantaging specific groups, ensuring that judgements are equitable across diverse subgroups.

3. Individual Fairness: AI systems should treat persons with similar features or actions identically, fostering consistency in choices.

4. No Discrimination: AI systems should not make predictions or choices that discriminate against any group based on protected traits such as race, gender, or religion.

Strategies to Mitigate Bias and Enhance Fairness

Addressing bias and promoting fairness in deep learning systems needs a multi-pronged strategy that spans several stages of the AI development lifecycle:

1. Data Collection and Annotation: Start with broad and representative training data that mirrors the real-world population. Careful data annotation that reduces subjective biases is key.

2. Data Preprocessing: Employ approaches to detect and minimise bias in training data. Resampling, data augmentation, and oversampling can help balance data distributions.

3. Algorithmic Fairness: Develop methods that explicitly consider fairness during model training. Various fairness-aware strategies, such as adversarial debiasing and reweighting, can be applied.

4. Bias Auditing and Testing: Regularly audit and test AI models for biases using fairness measures and analysis tools. Identify and fix biases in forecasts and judgements.

5. Diverse Development Teams: Ensure that AI development teams are diverse and inclusive, representing multiple backgrounds, opinions, and experiences. This can help uncover and remove biases early in the development process.

6. Ethical principles and laws: Adhere to ethical principles and laws that promote justice and eliminate biases in AI systems. Regulatory agencies and industry norms play a significant role in ensuring responsible AI development.

Conclusion

The rise of deep learning has brought both revolutionary promise and ethical issues to the forefront. Bias and fairness issues in deep learning systems necessitate urgent attention and intelligent mitigating measures. As AI technologies become increasingly integrated in our lives, it's crucial to guarantee that these systems do not perpetuate inequities, prejudice, or unfairness. By implementing diverse datasets, algorithmic fairness, ongoing monitoring, and adherence to ethical rules, developers can contribute to the building of AI systems that respect human values, uphold society norms, and actually empower individuals and communities. Building fair AI systems is not only a technological problem but also a moral obligation that dictates the direction of AI's impact on society.

2.3 Case Studies: Examining Ethical AI Failures

In the era of artificial intelligence, the promise of innovation and advancement is confronted with the obligation to ensuring that AI technologies adhere to ethical standards and societal values. However, the route towards ethical AI development is not without its perils, and history is marked by occasions where AI systems have failed to fulfil these requirements, leading to unexpected effects, biases, and ethical difficulties. In this chapter, we look into case studies that shed light on ethical AI failures, investigating the core causes, consequences, and the lessons we may take from these experiences to guide the responsible development of AI systems.

1. Tay, the Chatbot with Offensive Behavior

The instance of Microsoft's chatbot, Tay, serves as a sharp warning of how AI systems might be susceptible to harmful influences. Launched on Twitter in 2016, Tay was supposed to learn and engage in discussions with people, altering its vocabulary based on interactions. However, the system quickly became a forum for harmful and inappropriate content, as malevolent individuals used its learning capabilities to teach it inflammatory and discriminating words.

Lesson Learned: This scenario underlines the need of evaluating potential misuse and hostile behavior during AI system development. Robust content filtering techniques and ethical norms are required to avoid AI systems from amplifying bad information.

2. COMPAS Algorithm and Criminal Sentencing Bias

The story of the COMPAS (Correctional Offender Management Profiling for Alternative Sanctions) algorithm demonstrates the ethical challenges of algorithmic decision-making in the criminal justice system. The algorithm is designed to assess the likelihood of recidivism among individuals to guide decisions about bail, punishment, and parole. However, investigations have indicated that the algorithm demonstrates racial bias, disproportionately identifying African American defendants as high risk and potentially contributing to biassed sentence decisions.

Lesson Learned: Bias can be built in algorithmic decision-making processes, leading to discriminatory consequences. This case underlines the significance of constant monitoring, transparency, and accountability in maintaining fairness and equity in AI systems employed in vital fields like criminal justice.

3. Amazon's Biased Recruitment Algorithm

In an effort to improve its recruitment process, Amazon built an AI-based hiring tool to evaluate resumes and find possible employees. However, the algorithm demonstrated gender bias, favoring male candidates over female ones. The prejudice developed because the model was trained on historical resumes, which were weighted towards male applicants due to gender imbalances in the computer industry.

Lesson Learned: The bias contained in historical data can be perpetuated by AI systems, leading to biassed consequences. This scenario underlines the need for diverse and representative training data and proper data preprocessing to prevent such biases from appearing.

4. Google Photos' Offensive Labels

In 2015, Google Photos attracted outrage after its AI-driven auto-tagging function branded a photo of two African American individuals as "gorillas." This incident demonstrated the limitations of AI systems' ability to analyse and categorize complex visual data, thus perpetuating racial preconceptions.

Lesson Learned: This case underscores the significance of thorough testing and quality assurance in AI systems, particularly when dealing with sensitive content. Developers must anticipate potential inaccuracies and biases in forecasts to prevent undesirable outcomes.

5. Facial Recognition and Privacy Concerns

Facial recognition technology has prompted substantial ethical problems, notably relating to privacy and monitoring. Clearview AI, a startup that built a facial recognition system, gathered billions of photographs from the internet to construct a database for police enforcement organisations. This sparked worries about the absence of consent, the potential for misuse, and the degradation of privacy.

Lesson Learned: The collecting and use of personal data for AI systems must adhere to strong ethical rules and legislation. Transparency, informed permission, and clear reasons for data usage are crucial to defend persons' privacy rights.

Conclusion

The investigation of ethical AI failures serves as a sobering reminder that despite the promise of AI technology, there are inherent hazards that demand care and responsible development. These case studies underscore the necessity for interdisciplinary collaboration, ethical principles, transparency, and comprehensive testing throughout the AI development lifecycle. Understanding the core causes of ethical failures gives useful insights that drive the mitigation of biases, prevention of misuse, and alignment of AI systems with human values.

As AI technologies continue to improve, the lessons from these case studies underline the importance of ethical considerations in AI development. By learning from past mistakes, engaging in ongoing ethical reflection, and prioritizing fairness, transparency, and accountability, developers can contribute to a future where AI systems are not only innovative and capable but also ethically sound and respectful of the rights and dignity of all individuals and communities.

Chapter 3
Explainability Techniques

3.1 The Importance of AI Explainability

In the field of artificial intelligence (AI), the growth of complex and sophisticated models has offered unparalleled possibilities, from diagnosing diseases to autonomous driving. However, this advancement has also led to a significant challenge: the lack of openness and interpretability in AI systems. The concept of AI explainability has arisen as a fundamental component of responsible AI development, demanding that AI models not only produce accurate predictions but also provide clear and relevant explanations for their judgements. This chapter looks into the relevance of AI explainability, analysing its consequences, benefits, obstacles, and the ethical obligation to guarantee that AI systems are transparent and responsible to human comprehension.

Understanding AI Explainability

AI explainability refers to the ability of an AI system to deliver clear and intelligible explanations for its actions and forecasts. It includes making the internal workings of complicated AI models comprehensible to humans, letting users to comprehend how the model arrives at a given outcome. Explainability goes beyond black-box models, which make forecasts without revealing the underlying processes, to models that can express the rationale behind their conclusions.

The Implications of Lack of Explainability

The absence of AI explainability has substantial ramifications for numerous stakeholders, including developers, consumers, regulators, and society as a whole:

1. Lack of Trust: When users cannot grasp how AI systems arrive at judgements, trust in these systems erodes. This lack of confidence can limit adoption and acceptance of AI technologies.

2. Accountability: In fields such as healthcare and finance, where AI systems make vital judgements, accountability is essential. If judgements are not explainable, it's tough to hold developers and organizations responsible for errors or prejudices.

3. Bias and Fairness: Lack of explainability can mask biases present in AI systems. Understanding how biases are introduced and propagated is vital for addressing fairness concerns.

4. Ethical Concerns: The ethical consequences of AI judgements become more obvious when they cannot be explained. In high-stakes settings like medical diagnostics or criminal punishment, the ethical underpinning for choices must be apparent.

Benefits of AI Explainability

AI explainability offers a range of benefits that extend beyond mere transparency:

1. User Empowerment: Explainable AI empowers users to make educated decisions based on the model's explanations. Users can trust the system and have the ability to remedy problems or question judgements.

2. Error Detection and Correction: With explanations, users can detect flaws, misinterpretations, or biases in the model's thinking. This feedback loop aids in model improvement and refining.

3. Bias Detection and Mitigation: Explainability enables for the detection of biases in model decisions, enabling developers to address biases and promote fairness.

4. Regulatory Compliance: In businesses subject to rules, explainability is vital for compliance and auditability. Regulators need to ensure that AI systems make decisions based on reasonable and justifiable thinking.

Challenges in Achieving AI Explainability

Achieving AI explainability is not without challenges:

1. Complex Models: Deep neural networks and other complex models are sometimes considered "black boxes" due to their intricate internal workings. Explaining their decisions can be tough.

2. Trade-off with Performance: Adding explainability features may trade off with model performance. Simplified models might not reach the same accuracy as complicated ones.

3. Balancing Simplicity and Interpretability: Making models interpretable could necessitate simplification, which might diminish accuracy. Balancing accuracy and interpretability is a delicate task.

4. Domain Complexity: In complicated fields like medical diagnosis or financial analysis, the underlying processes are intricate and not immediately translatable into human-readable explanations.

Ethical Imperative of AI Explainability

Beyond technical reasons, the ethical requirement for AI explainability is undeniable:

1. Human Autonomy: In fields affecting human life, humans have the right to comprehend the foundation of AI decisions that impact them. AI technologies should enhance human judgment, not replace it.

2. Preventing Discrimination: Explainability aids in recognising and moderating biases, ensuring that AI systems do not perpetuate or magnify unfair discrimination.

3. Accountability and Responsibility: Developers and organizations must be accountable for the consequences of their AI systems. Explainability is a critical step towards fulfilling this commitment.

4. Transparency in AI Governance: Societies should have transparent and intelligible AI systems, enabling accountability, trust, and responsible AI governance.

Several ideas and procedures can assist make AI systems more explainable:

1. Feature Importance: Identifying elements that contribute most to a choice helps consumers understand which aspects influence the outcome.

2. Local Explanations: Explaining decisions on a per-instance level, showing how a given input led to a certain prediction.

3. Rule-Based Models: Using decision trees or rule-based models, which provide a sequence of logical rules for decisions.

4. Saliency Maps: Visualizing regions of input data that significantly influence predictions, helping users understand model emphasis.

5. Counterfactual Explanations: Providing alternate inputs that would lead to different outcomes, enhancing understanding of decision boundaries.

Conclusion

In a society increasingly driven by AI technology, AI explainability is not simply a technological challenge but a fundamental ethical imperative. The ability of AI systems to deliver clear, understandable, and relevant explanations for their judgements is vital for transparency, accountability, and confidence. AI developers, academics, policymakers, and society at large must collaborate to guarantee that AI technologies match with human values, empower users, and promote fairness and ethical decision-making. The path to responsible AI development must be paved with explainability, enabling us to embrace the potential of AI while maintaining human understanding and control over its decisions.

3.2 Local vs. Global Explainability: Unveiling the Scope of AI Insights

As the field of artificial intelligence (AI) continues to grow, the necessity for explainability has become increasingly clear. AI systems are already capable of intricate decision-making, yet their inner workings typically remain unintelligible to humans. This has led the examination of diverse approaches to explainability, with the distinction between local and global explainability emerging as a significant factor. This chapter goes into the intricacies of local and global explainability, analysing their definitions, applications, advantages, and limitations, and shining light on how these techniques might strengthen our knowledge of AI systems and their decision-making processes.

Defining Local and Global Explainability

Local Explainability: Local explainability refers to the ability to grasp the rationale behind a specific choice made by an AI model for a particular instance of input data. It focuses on offering insights into why a model arrived at a given prediction or outcome for a single input.

Global Explainability: In contrast, global explainability incorporates an understanding of how an AI model operates throughout the entire dataset or a specific area. It seeks to disclose the overall patterns, correlations, and properties that the model has learned from the entire dataset.

Applications of Local Explainability

Local explainability is very beneficial when we want to understand why a certain AI decision was made for a given input instance:

1. Medical Diagnosis: In healthcare, local explainability can assist doctors comprehend why a certain diagnosis was made for a patient, providing insight into the underlying elements.

2. Credit Scoring: For credit applications, local explainability helps highlight why an applicant was granted or denied credit, letting others appreciate the elements that drove the decision.

3. Autonomous Driving: In self-driving automobiles, local explainability helps elucidate why the vehicle made a certain maneuver, allowing passengers to trust and comprehend the vehicle's behaviour.

Advantages and Limitations of Local Explainability

Advantages:

1. Interpretable Insights: Local explainability provides clear insights into why a given decision was made, bringing transparency and clarity.

2. User Empowerment: Users can fix errors, debate conclusions, or learn from AI's thinking, contributing to a more informed and empowered user base.

3. Feedback Loop: Local explanations promote a feedback loop for model improvement, since users can provide input on specific situations when the model might have made errors or biased conclusions.

Limitations:

1. Limited Context: Local explanations are confined to a single incident and may not represent broader patterns or trends contained in the data.

2. Resource Intensive: Generating local explanations can be computationally expensive, especially for complicated models and huge datasets.

Applications of Global Explainability

Global explainability strives to identify insights that apply across a broader scope, enabling a more comprehensive understanding of AI models:

1. Algorithm Evaluation: Researchers can utilise global explanations to evaluate the overall behavior of multiple AI algorithms on a dataset, comparing their strengths and faults.

2. Fairness Analysis: Global explanations can reveal patterns of bias or fairness concerns that might be common across the entire dataset, contributing in the development of fairer models.

3. Feature Importance: By examining global explanations, researchers may discover the features that regularly contribute to model predictions, helping to understand the driving causes.

Advantages and Limitations of Global Explainability

Advantages:

1. Holistic Understanding: Global explanations offer insights into the entire behavior of AI models, enabling consumers and developers understand how models make judgements across multiple circumstances.

2. Detecting Bias: Global explanations can uncover systematic biases inherent in the dataset, leading to more equitable model development.

3. Insights for Model Design: Developers can receive insights about the most significant features and relationships, driving the design of future models.

Limitations:

1. Lack of Specificity: While global explanations provide overall insights, they may not provide precise insights into why a particular decision was made for a specific case.

2. Complexity: Understanding global behavior can be tough, especially for complicated models with intricate interactions and non-linear correlations.

Balancing Local and Global Explainability

Rather than an either-or strategy, local and global explainability can complement each other, enabling a more complete knowledge of AI systems. Local explanations provide instant insights into specific decisions, whereas global explanations assist discover patterns, trends, and biases that affect the system as a whole.

Conclusion

The distinction between local and global explainability constitutes a critical consideration in the creation of transparent and accountable AI systems. Each technique offers unique insights and benefits, and their interplay can expand our understanding of AI models' decision-making processes. By applying a balanced approach to explainability, developers, researchers, and users may work towards AI systems that not only operate successfully but also deliver valuable insights that correspond with human values and society standards. Whether striving to understand a single choice or revealing bigger trends, the pursuit of both local and global explainability is a vital step toward responsible and ethical AI development.

3.3 Interpretable Models: LIME, SHAP, and More

The intricacy of modern machine learning models, such as deep neural networks, has ushered in amazing developments in artificial intelligence. However, this intricacy typically comes at the cost of interpretability, leaving these models as "black boxes" that provide accurate forecasts but lack transparency in their decision-making. The necessity for interpretable models has led to the creation of numerous methodologies, including Local Interpretable Model-agnostic Explanations (LIME) and Shapley Additive Explanations (SHAP), among others. This chapter digs into these strategies and more, studying how they bridge the gap between model complexity and human understanding, revealing insights into the inner workings of AI systems.

The Need for Interpretability

As AI models grow vital to critical decision-making in domains like healthcare, finance, and law, the capacity to explain their decisions becomes paramount. Black-box models can yield accurate answers, but their inability to provide meaningful explanations affects user trust, accountability, and the discovery of biases. Interpretability is vital to guarantee that AI models conform with ethical and legal standards, encourage transparency, and empower users to comprehend, validate, and potentially correct the decisions made by these models.

Local Interpretable Model-agnostic Explanations (LIME)

LIME is a popular strategy for explaining the predictions of complex models by approximating them with simple, interpretable models that are easier to understand. LIME functions on a local level, providing explanations for particular predictions rather than the entire model.

How LIME Works:

1. Selecting an Instance: LIME starts by selecting a specific instance for which the model's prediction needs to be explained.

2. Generating Perturbations: The selected instance is perturbed to build a collection of comparable examples with small changes.

3. Model Approximation: A simpler, interpretable model (e.g., linear regression) is trained on the altered dataset to approximate the behavior of the complicated model for that instance.

4. Interpretable Explanation: The interpretable model's coefficients are used to explain the prediction, demonstrating the importance of each characteristic to the forecast.

Advantages of LIME:

1. Model-Agnostic: LIME works with any machine learning model, regardless of its complexity or kind.

2. Local Explanations: LIME provides explanations for specific situations, delivering context-sensitive insights.

3. Simplicity: LIME's explanations are straightforward to understand, especially for non-experts.

Shapley Additive Explanations (SHAP)

SHAP is anchored in cooperative game theory and gives a coherent framework for analysing model predictions. It provides a consistent mechanism to credit the contribution of each feature to a prediction, both at the individual level and throughout the entire dataset.

How SHAP Works:

1. Shapley Values: SHAP allocates a value to each characteristic that represents its contribution to the prediction, following the concept of Shapley values in game theory.

2. Explaining Predictions: SHAP values are used to explain individual predictions by demonstrating how far each attribute pushes the prediction away from the average prediction.

3. Global Explanations: By aggregating SHAP values throughout the entire dataset, SHAP provides insights on feature relevance and model behavior on a global scale.

Advantages of SHAP:

1. Consistency: SHAP gives a unifying explanation structure that is consistent across distinct prediction explanations.

2. Global and Local Insights: SHAP gives both individual-level and global insights, allowing users to comprehend overall trends and specific predictions.

3. Feature Importance: SHAP quantifies the contribution of each feature to a prediction, facilitating in feature selection and interpretation.

Other Techniques for Interpretability

1. Decision Trees: Decision trees are intrinsically interpretable, presenting a set of decisions and rules that lead to a forecast. However, they could lack the predictive potential of more complicated models.

2. Partial Dependence Plots (PDP): PDPs depict how a given feature effects predictions while keeping other features constant.

3. Feature Importance Scores: Techniques like permutation feature importance and feature importance from tree-based models provide insights into which characteristics are most influential.

Challenges and Considerations

While interpretable approaches offer valuable insights, there are issues to consider:

1. Accuracy-Interpretability Trade-off: Interpretable models could compromise some predicted accuracy for the sake of explainability.

2. Complexity: Balancing interpretability with the complexity of modern AI models is a tough task.

3. Domain Knowledge: Interpretability typically requires domain experience to generate meaningful conclusions from explanations.

Conclusion

The hunt for interpretable models is a critical step in bridging the gap between complicated AI systems and human understanding. Techniques like LIME, SHAP, decision trees, and feature relevance scores provide vital insights into how AI models arrive at their predictions. These methodologies empower users, regulators, and developers to evaluate, validate, and enhance AI systems, ensuring that they operate transparently, ethically, and accountably. In a world where AI decisions touch individuals and society at large, interpretable models offer a fundamental foundation for ethical AI development.

3.4 Transparency in Deep Learning Models: Illuminating the Black Box

Deep learning models have showed impressive skills across numerous areas, from image identification to natural language processing. However, their increasing complexity and the inherent opacity of their decision-making processes have given rise to issues about accountability, fairness, and trust. The concept of transparency in deep learning models has developed as a significant initiative to shed light on these "black box" systems, enabling humans to comprehend, validate, and interpret their predictions. In this chapter, we delve into the significance of transparency in deep learning models, investigating the problems, benefits, approaches, and ethical issues connected with making these models more accessible and responsible.

Understanding Transparency in Deep Learning Models

Transparency refers to the ability to understand the inner workings of a system, comprehend its decision-making processes, and have access to meaningful explanations for its actions. In the context of deep learning models, transparency tries to disclose how these complicated networks arrive at their predictions, allowing users to get insights into the elements and attributes that impact their conclusions.

Challenges in Deep Learning Transparency

Achieving transparency in deep learning models is not without challenges:

1. Complexity: Deep neural networks can have millions of parameters and sophisticated structures, making it tough to trace the cause-and-effect links that lead to precise predictions.

2. Non-Linearity: Deep learning models often display non-linear interactions between features, making it difficult to intuitively understand how changes in input effect output.

3. High-Dimensional Data: Many applications of deep learning, such as picture and speech recognition, include high-dimensional data, which is tough to interpret in a human-understandable manner.

Benefits of Transparency in Deep Learning Models

Transparency in deep learning models offers a range of benefits:

1. Trust and Accountability: Transparent models inspire trust by allowing users to evaluate the correctness of forecasts. They also enhance accountability by enabling developers to analyse and remedy problems.

2. Bias Detection and Mitigation: Transparent models promote the discovery of biases in training data, algorithms, and decision processes, leading to fairer and more equitable outcomes.

3. Regulatory Compliance: In regulated businesses, transparent models ensure that AI systems adhere to legal and ethical norms, simplifying auditability and compliance.

4. User Empowerment: Transparency empowers users to comprehend AI decisions, supporting informed decision-making and enabling users to question or rectify wrong predictions.

Techniques for Achieving Transparency in Deep Learning Models

Several strategies have been explored to promote transparency in deep learning models:

1. Interpretable Models: Utilizing simpler models like decision trees or linear regression to imitate the behavior of larger deep learning models, providing intelligible explanations.

2. Local Explanation Methods: Techniques like Local Interpretable Model-agnostic Explanations (LIME) produce explanations for individual predictions by approximating the model's behavior around specific cases.

3. Feature Visualization: Visualizing which elements of an input contribute most to a given prediction, aiding in understanding the model's focus and decision process.

4. Attention Mechanisms: Attention mechanisms emphasise which elements of the input the model focuses on during prediction, revealing insights into its decision-making.

5. Feature value: Quantifying the value of input characteristics in impacting model predictions, helps in understanding the influence of each feature.

Ethical Implications of Transparency in Deep Learning Models

Transparency in deep learning models brings ethical implications:

1. Bias and Fairness: Transparent models enable the detection and mitigation of biases, ensuring that AI systems are fair and do not perpetuate discrimination.

2. Human Dignity: Users have the right to comprehend the rationale of AI decisions that influence their life, respecting their dignity and individuality.

3. Accountability: Developers and organizations are accountable for the consequences of their AI systems. Transparent models enable insights to fulfill this commitment.

4. private: Transparency must be achieved while protecting individuals' private rights, ensuring that sensitive data is not revealed during explanation processes.

Balancing Transparency with Performance

There is a trade-off between transparency and model performance:

1. Accuracy: Making models more transparent could include simplifications that impair accuracy. Striking the appropriate balance is key.

2. Complexity: Deep learning models are inherently complex, and establishing high degrees of openness could be tough without losing predictive power.

Conclusion

In an age where AI systems increasingly affect human lives, the goal of transparency in deep learning models is an important endeavor. Transparency empowers users, promotes justice, fosters accountability, and upholds ethical norms. While attaining transparency in large deep learning models offers obstacles, the benefits are far-reaching. Through techniques like interpretable models, local explanations, and feature visualization, developers may navigate the balance between complexity and transparency, ensuring that AI systems are not just strong but also accessible and aligned with human values. As we continue to incorporate AI into our society, the desire for transparent AI models becomes even more critical, building a future where technology empowers, informs, and respects individuals and communities.

Chapter 4

Interpretable Neural Networks

4.1   Visualizing Convolutional Neural Networks (CNNs): Unveiling Deep Learning Insights

Convolutional Neural Networks (CNNs) have transformed the science of computer vision, enabling machines to receive and interpret visual data with human-like precision. Despite their extraordinary performance, CNNs frequently operate as sophisticated "black box" systems, making it tough to grasp how they arrive at their predictions. The concept of visualizing CNNs has evolved as a useful tool to illuminate the inner workings of these networks, enabling insights into feature extraction, hierarchical representation, and decision-making processes. In this chapter, we delve into the relevance of visualizing CNNs, studying the methodology, applications, benefits, and ethical considerations connected with getting a deeper understanding of these sophisticated models.

Importance of Visualizing CNNs

Understanding how CNNs analyse and interpret visual data is crucial for various reasons:

1. Model Validation: Visualizing CNNs helps confirm that they are learning significant features from the data, boosting confidence in their performance.

2. Bias Detection: Visualization can identify biases or artifacts in the learnt features, allowing developers to fix unwanted biases in the model's behavior.

3. Feature Interpretation: By displaying the characteristics recognised by multiple layers of the CNN, researchers may interpret what the network is learning at different levels of abstraction.

Techniques for Visualizing CNNs

Several ways have been developed to visualize the inner workings of CNNs:

1. Activation Maps: Activation maps emphasise the portions of an image that activate specific filters in a CNN, exposing the qualities that excite certain neurons.

2. Filter Visualization: Techniques like Gradient Ascent change an input image to optimise the activation of a single neuron, resulting in a visual depiction of what the filter is detecting.

3. Class Activation Maps (CAM): CAM emphasises the parts of an image that contribute most to a given class prediction, helpful in understanding the model's decision process.

4. Feature Visualization: Techniques like DeepDream generate bizarre pictures that enhance the activation of certain features, providing insight into what the network finds intriguing.

Benefits of Visualizing CNNs

Visualizing CNNs gives a plethora of benefits:

1. Transparency: Visualization demystifies the decision-making process, boosting the transparency of CNNs and making their predictions more understandable.

2. Interpretability: Researchers and users can understand the visible features to get insights into how the CNN recognises and categorizes items.

3. Model Improvement: Visualization can assist discover shortcomings in CNNs, helping engineers to fine-tune models for greater performance.

Applications of Visualizing CNNs

1. Medical Imaging: Visualizing CNNs can benefit radiologists in understanding how the network discovers anomalies in medical pictures, leading to enhanced diagnosis accuracy.

2. Autonomous Vehicles: In self-driving automobiles, displaying CNNs can illustrate how the network identifies pedestrians, traffic signs, and other things, boosting safety.

3. Agriculture: CNN visualization can assist farmers in comprehending how the network recognises crop diseases or insect infestations, enabling early interventions.

Ethical Considerations in CNN Visualization

While visualizing CNNs is a great tool, it presents ethical considerations:

1. Privacy: Visualized features could reveal sensitive information about individuals or locations in photographs, necessitating careful handling of data.

2. Biases: Visualization can disclose biases in the data used for training, underlining the significance of addressing fairness and equity problems.

3. Misinterpretation: Overinterpretation of visible features might lead to inaccurate conclusions or misunderstanding of the model's behavior.

Challenges in Visualizing CNNs

1. High-Dimensional Data: Images are high-dimensional, and mapping CNN activations to interpretable visualizations can be problematic.

2. Complex Architectures: CNNs frequently consist of numerous layers with intricate interactions, making it difficult to determine the specific role of each layer.

3. Non-Linearity: The non-linear transformations done by CNN layers challenge the interpretation of how input features translate into higher-level representations.

Conclusion

In the era of AI, understanding and interpreting the decisions of complicated models like CNNs is vital for accountability, trust, and ethical deployment. Visualizing CNNs bridges the gap between their black-box nature and human comprehension, bringing insights into feature extraction, hierarchical representation, and decision-making processes. Through approaches like activation maps, filter visualization, and class activation maps, developers and researchers may reveal the mysteries of these networks, enabling us to harness their potential while ensuring that their behaviors correspond with human values and ethical considerations. As CNNs continue to alter industries and impact lives, the ability to view and explain their operations becomes a vital tool for establishing a more transparent, educated, and ethically sound AI ecosystem.

4.2   Attention Mechanisms: Making Neural Networks Transparent

In the domain of artificial intelligence, the advent of large neural networks has led to revolutionary advancements in tasks like natural language processing, picture identification, and more. However, these complex models frequently operate as "black boxes," making it impossible to understand how they arrive at their judgements. The emergence of attention mechanisms has emerged as a transformative tool to bring transparency to these complicated brain networks. By allowing networks to focus on relevant information, attention mechanisms provide insights into the inner workings of models, making them more interpretable, accountable, and human-like in their decision-making. In this chapter, we delve into the significance of attention mechanisms, investigating their applications, benefits, methodologies, and the ways they help to making neural networks transparent.

Understanding Attention Mechanisms

At its foundation, attention refers to the ability to selectively focus on specific portions of input material while disregarding others. Attention mechanisms in neural networks replicate this cognitive function by enabling models to assign various weights to different regions of the input, raising the value of important information and lowering the influence of irrelevant data. This technique helps the model to make decisions while emphasizing the most essential elements.

Applications of Attention Mechanisms

Attention techniques have transformed numerous domains:

1. Natural Language Processing: In machine translation, attention mechanisms allow models to focus on specific words in the source sentence while creating the target translation, resulting in more accurate and contextually relevant translations.

2. Image Captioning: In image captioning challenges, attention methods enable models to attend to diverse sections of the image while creating descriptive captions, increasing the coherence of the output text.

3. Question Answering: Attention mechanisms boost question-answering models by helping them to focus on relevant areas of the input text when generating answers, resulting to more correct responses.

Benefits of Attention Mechanisms

1. Interpretability: Attention methods provide a clear indicator of which sections of the input data the model is focusing on, making the decision-making process more interpretable.

2. Transparency: By revealing the attention weights assigned to distinct input items, attention mechanisms contribute to the transparency of neural network predictions.

3. Local and Global Context: Attention mechanisms enable models to incorporate both local and global context, ensuring that decisions are made in a more informed manner.

4. Bias Mitigation: Attention methods can assist minimise biases by allowing models to focus on key characteristics independent of inherent biases in the data.

Techniques for Implementing Attention Mechanisms

Several strategies are employed to include attention mechanisms into neural networks:

1. Self-Attention: Also known as intra-attention or scaled dot-product attention, this technique computes attention weights based on the similarity between distinct pieces of the input data.

2. Multi-Head Attention: This strategy uses multiple sets of attention weights to capture distinct parts of the input, boosting the model's ability to grasp complicated interactions.

3. Visual Attention: Applied in image processing tasks, visual attention mechanisms enable models to focus on different portions of an image while generating descriptive text.

Ethical Considerations in Attention Mechanisms

As with any AI technology, attention processes carry ethical implications:

1. Bias: If attention processes focus on biased aspects of the input data, they can perpetuate existing prejudices or discriminate against particular groups.

2. Transparency: While attention processes provide insights, overreliance on them can produce a false sense of comprehension, potentially leading to inaccurate conclusions.

3. Interpretation: Visualizations of attention weights can be complex and tough to understand appropriately, requiring domain expertise for valuable insights.

Challenges and Future Directions

1. Complexity: Implementing attention mechanisms can increase the complexity of neural network topologies, increasing training time and resource needs.

2. Interpretability: While attention processes provide insights, they could not necessarily lead to a thorough understanding of the model's decision-making, especially in highly complicated tasks.

3. Hyperparameter adjustment: The effectiveness of attention mechanisms can be modified by hyperparameters that require careful adjustment.

Conclusion

Attention processes constitute a critical step in making massive neural networks transparent, responsible, and interpretable. By allowing models to focus on relevant information, attention mechanisms give insights into decision-making processes that were once unfathomable. Their applications in natural language processing, image processing, and other domains have revolutionised how AI systems work, boosting their performance and ethical considerations. As the area of AI continues to evolve, attention processes are positioned to play a vital role in enabling models to imitate human cognitive abilities, promoting trust, and driving ethical AI development. By shedding light on the inner workings of neural networks, attention mechanisms open the path for a more informed and transparent future in AI.

4.3 Capturing Feature Importance with Grad-CAM: A Window into Deep Learning Insights

The rapid growth of deep learning has transformed several sectors, from image identification to medical diagnosis. However, the intricacy of deep neural networks often makes it tough to explain why these models arrive at specific conclusions. Grad-CAM (Gradient-weighted Class Activation Mapping) has emerged as a pioneering technique that allows a view into the black box, providing insights into which components of an input contribute most to a model's conclusion. This chapter goes into the significance of Grad-CAM, studying its applications, benefits, implementation, and how it increases our understanding of deep learning models by capturing feature importance.

Understanding Grad-CAM

Grad-CAM is a technique aimed to show the portions of an input that are most relevant for a deep learning model's prediction. By highlighting these locations, Grad-CAM presents an explanation for why a certain prediction was produced, reducing complex forecasts into interpretable insights. This method is particularly powerful for convolutional neural networks (CNNs) utilised in image-related applications.

Applications of Grad-CAM

Grad-CAM finds applications in a wide range of fields:

1. Medical Imaging: In medical diagnosis, Grad-CAM can highlight the portions of an image that influenced a model's prediction, helping clinicians understand why a given diagnosis was reached.

2. Object Detection: For object detection tasks, Grad-CAM can pinpoint the areas of an image that contributed most to the classification of an object, enhancing the interpretability of detection models.

3. Autonomous Vehicles: In self-driving automobiles, Grad-CAM can illustrate why the model made specific decisions, providing insights into the detection of pedestrians, vehicles, and traffic signs.

Benefits of Grad-CAM

Grad-CAM offers various advantages:

1. Interpretability: By displaying the parts of an image that influenced the prediction, Grad-CAM makes the decision-making process of deep learning models more interpretable.

2. Transparency: Grad-CAM enhances the transparency of model predictions, allowing users to understand why a given forecast was produced.

3. Model Validation: Grad-CAM helps confirm that the model is focused on essential features, boosting the model's credibility and trustworthiness.

The application of Grad-CAM comprises multiple steps:

1. Select a Target Layer: The first step is to select a layer within the model to investigate. Typically, this layer should be close to the output, where features are already abstracted.

2. Compute Gradients: Gradients of the anticipated class score with respect to the feature maps of the selected layer are computed. These gradients demonstrate how much each feature map contributes to the final prediction.

3. Pooling of Gradients: The gradients are pooled to obtain the importance of each feature map. This pooling is frequently done by taking the average of the gradients.

4. Weighted Sum of Feature Maps: The pooled gradients are multiplied with the feature maps of the selected layer. This stage emphasises the parts of the feature maps that contributed most to the prediction.

5. ReLU and Upsampling: Applying a ReLU (Rectified Linear Unit) to the weighted feature maps guarantees that only positive contributions are examined. The generated feature maps are then upsampled to fit the proportions of the original image.

6. Visualization: The upsampled feature maps are superimposed on the original image, graphically highlighting the regions that are essential for the model's prediction.

Ethical Considerations in Grad-CAM

As with any visualization approach, Grad-CAM involves ethical considerations:

1. Privacy: Visualized heatmaps could mistakenly reveal sensitive information in photographs, stressing the necessity to secure data privacy and security.

2. Bias: If the training data contains biases, Grad-CAM can possibly magnify these biases, promoting unjust or unfair conclusions.

3. Misinterpretation: Visualizations can be misconstrued, leading to inaccurate assumptions about the model's behavior.

Challenges and Future Directions

1. High-Dimensional Data: Visualizing feature relevance in high-dimensional images can be tricky, requiring careful processing and presentation.

2. Complex topologies: Implementing Grad-CAM for models with intricate topologies or multiple branches requires adjustments to ensure correct insights.

Conclusion

In the drive to make deep learning models more visible and interpretable, Grad-CAM stands as a powerful tool that exposes the decision-making process. By highlighting the parts of an input image that are critical for a model's prediction, Grad-CAM turns complex models into more understandable systems. Its applications in medical imaging, object identification, and autonomous cars illustrate its adaptability across fields. As deep learning continues to advance and touch numerous industries, tools like Grad-CAM pave the way for responsible AI development. By bridging the gap between sophisticated models and human comprehension, Grad-CAM gives a window into the black box, enabling users, developers, and researchers to test, analyse, and trust the judgements produced by deep learning models.

Chapter 5

Ethics in Data Collection and Preprocessing

5.1   Addressing Bias in Training Data: Ensuring Fair and Ethical AI

In the area of artificial intelligence (AI), training data serves as the basis upon which machine learning models are developed. However, this data is not immune to biases prevalent in the actual world, sometimes leading to flawed AI models that perpetuate and even magnify unfair imbalances. The need of reducing bias in training data has never been more vital, as AI systems continue to play an increasingly significant role in decision-making across numerous sectors. This chapter digs into the significance of correcting bias in training data, analysing the problems, methodologies, ethical concerns, and the broader impact of assuring justice and equity in AI systems.

Understanding Bias in Training Data

Bias in training data refers to the presence of systematic and unfair favouring or prejudice toward specific groups or features. This bias can come from several origins, including historical imbalances, societal prejudices, or data gathering techniques that unwittingly increase existing inequalities.

Challenges Posed by Bias in Training Data

Addressing bias in training data raises various challenges:

1. Unconscious Bias: Biases in data can be subtle and unconscious, making them difficult to detect and control.

2. Data Availability: Certain groups might be underrepresented in the training data, leading to poor model performance for such groups.

3. Algorithmic Reinforcement: Biased training data can lead to biased AI models that reinforce existing inequities, worsening the problem.

Methods to Address Bias in Training Data

1. Diverse and Representative Data: Ensuring that training data is diverse and representative of the population being modeled can decrease bias by offering a balanced view of different groups.

2. Data Augmentation: Synthetic data can be developed to balance underrepresented groups, decreasing the bias associated with scarce data.

3. Bias Detection Tools: Employing specialist tools to identify and quantify bias in training data helps developers grasp the magnitude of the problem.

4. Preprocessing Techniques: Techniques like re-sampling or re-weighting can balance the representation of different groups in the training data.

Ethical Implications of Addressing Bias

Addressing bias in training data coincides with ethical considerations in AI development:

1. Fairness: Fairness in AI is a crucial ethical principle. Ensuring unbiased training data ensures equal outcomes for all individuals and groups.

2. Accountability: Developers and organizations are accountable for the impact of their AI systems. Addressing bias indicates accountability and respect to ethical principles.

3. Inclusivity: Addressing bias guarantees that AI systems are inclusive and do not discriminate against any population.

The Broader Impact of Addressing Bias

1. Social Equity: Addressing bias in training data contributes to a more just and equitable society by avoiding AI systems from replicating historical imbalances.

2. Trust and Acceptance: Unbiased AI models are more likely to be trusted and accepted by users, leading to greater adoption and beneficial societal impact.

3. creativity: By minimising bias, AI developers make room for creativity that caters to varied demands and experiences.

Ongoing Efforts and Challenges

1. Data Collection Challenges: Collecting fair and representative data can be problematic, especially for underrepresented or marginalized populations.

2. Algorithmic Complexity: Mitigating bias while preserving model performance involves sophisticated strategies that balance fairness and accuracy.

3. Measurement and Metrics: Developing metrics to objectively quantify bias and fairness is an ongoing task that entails complex trade-offs.

Conclusion

Addressing bias in training data is not only a technical task; it is a moral responsibility. As AI becomes increasingly integrated in our lives, the potential for bias to reinforce disparities grows. By ensuring that training data is diverse, representative, and unbiased, AI engineers can contribute to fairer, more accountable, and ethical AI systems. The road to eliminating prejudice in AI is ongoing, requiring teamwork, awareness, and a commitment to social responsibility. As we aspire for a future where AI enhances human lives without perpetuating harm, eliminating bias in training data stands as a vital step toward designing AI systems that uphold ideals of fairness, equity, and inclusivity.

5.2 Data Augmentation and Fairness: Balancing Enhancement with Equity

In the realm of machine learning, data augmentation has emerged as a viable strategy to boost model performance and prevent overfitting. By artificially increasing the training dataset through various changes, data augmentation enriches the learning process, enabling models to generalize better. However, as AI applications become crucial to decision-making across multiple areas, the ethical implications of data augmentation in respect to fairness have gained significance. This chapter addresses the dynamic interplay between data augmentation and fairness, investigating how these two notions overlap, the issues they offer, and the solutions to balance enhancement with equity in AI systems.

Understanding Data Augmentation

Data augmentation involves applying a range of alterations to the original dataset, creating new instances with modest variances while keeping the basic qualities of the data. These transformations can include rotations, translations, flips, and adjustments in color and contrast, among others. Data augmentation is particularly helpful in settings with little training data, avoiding models from memorizing examples and boosting better generalization.

The Role of Data Augmentation in Fairness

Data augmentation, while boosting model performance, can accidentally accentuate or alleviate biases existing in the original training data. Addressing justice demands evaluating how these shifts affect different demographic groups. If particular groups are disproportionately represented in the training data, data augmentation could further skew the distribution and produce disparities in predictions.

Challenges at the Intersection of Data Augmentation and Fairness

1. Bias Amplification: Data augmentation techniques might amplify existing biases in the training data, leading to biased predictions that disproportionately affect particular populations.

2. Biased Augmentation: Choosing augmentation procedures that accidentally favor one group over another might bring new forms of prejudice.

3. Measurement of Fairness: Defining and quantifying fairness in the context of data augmentation is hard, as different augmentation procedures might have differing consequences on fairness indicators.

Strategies for Balancing Data Augmentation and Fairness

1. Equitable Augmentation: Ensuring that data augmentation techniques are performed evenly across different demographic groups can avoid biased amplification.

2. Fairness-Aware Augmentation: Developing strategies that take fairness into account during augmentation can lead to more equitable distributions of augmented data.

3. Fairness measures: Defining fairness measures unique to data augmentation helps guide developers in measuring and eliminating biases created by augmentation.

The Ethical Dimension of Data Augmentation and Fairness

1. Transparency: Being honest about data augmentation techniques and their possible impacts on justice encourages responsibility and trust in AI systems.

2. Inclusivity: Striving for fairness guarantees that AI systems do not discriminate against any population, supporting inclusivity and social equity.

3. Responsibility: Developers have a responsibility to examine and reduce the potential for data augmentation to reinforce biases, aligning with ethical AI development.

Benefits of Equitable Data Augmentation

1. Generalization: Equitable data augmentation can increase model generalization by offering a more balanced view of diverse demographic groups.

2. Bias Mitigation: Fair augmentation strategies can limit the amplification of existing biases, leading to more impartial AI systems.

3. Trust: Fair data augmentation promotes user trust by exhibiting a commitment to ethical and unbiased AI.

Conclusion

Data augmentation, a significant approach for increasing model performance, intersects with fairness in the area of AI. The ethical obligation of AI developers extends beyond accuracy to incorporate equality and diversity. Striking a balance between data enhancement and justice demands critical assessment of the potential biases induced by augmentation approaches. As AI systems grow increasingly important in moulding judgements across sectors including healthcare, finance, and law, maintaining fairness through data augmentation becomes paramount. By implementing equitable data augmentation approaches, AI engineers pave the road for responsible AI that not only achieves high accuracy but also preserves ideals of fairness, transparency, and societal impact. In the search of a more just and equitable AI landscape, data augmentation emerges as a cornerstone in constructing trustworthy and ethical AI systems.

5.3 Data Privacy and Consent: Balancing Innovation and Rights

In the digital age, data has become a cornerstone of innovation, enabling the development of advanced technologies like artificial intelligence (AI) and tailored services. However, the increased gathering, analysis, and utilization of personal data create serious concerns regarding privacy and individual rights. Data privacy and consent are basic ideas that play a key role in the growing environment of technology and society. This chapter digs into the subtle balance between data privacy, innovation, and individual rights, analysing the problems, ethical issues, rules, and the urgency to negotiate this delicate equilibrium.

Understanding Data Privacy and Consent

Data privacy refers to the control individuals have over their personal information, ensuring that it is gathered, processed, and disseminated in a manner that respects their rights. Consent, on the other hand, is the express authorization provided by individuals for organizations to gather and utilize personal data for particular reasons. Consent forms the cornerstone of data privacy, as it permits individuals to make informed decisions regarding their personal information.

Challenges in Balancing Innovation and Data Privacy

1. Rapid technical Advancements: The speed of technical progress often outpaces the establishment of appropriate rules, resulting to gaps in data privacy protection.

2. Data Proliferation: The expansion of data collecting from multiple sources, such as IoT devices, social media, and online activities, increases the danger of data misuse and breaches.

3. Granularity of Consent: Striking the correct balance between collecting thorough consent and not overwhelming consumers with extensive consent forms is a challenge.

Ethical Considerations in Data Privacy and Consent

1. Autonomy: Respecting individuals' autonomy and providing them control over their data is a crucial ethical factor in data privacy.

2. Transparency: Organizations must be upfront about how data will be used and ensure that users have a clear knowledge of the aims and consequences of data collecting.

3. Fairness: Data privacy and consent support the idea of fairness, barring the exploitation of individuals' data for unjust or discriminating reasons.

The Role of Regulations and Laws

1. General Data Protection Regulation (GDPR): GDPR, enacted by the European Union, is one of the most comprehensive data privacy legislation internationally. It empowers individuals with greater control over their data and puts severe responsibilities on corporations for data protection.

2. California Consumer Privacy Act (CCPA): This regulation offers Californian residents rights to know how their data is used, seek deletion of their data, and opt out of its sale.

3. Data Protection Laws in Other Jurisdictions: Countries around the world are implementing or amending data protection laws to address the issues brought by evolving data practices.

Strategies for Balancing Data Privacy, Consent, and Innovation

1. Privacy by Design: Organizations should build privacy issues into their technology and processes from the very beginning.

2. Granular Consent: Implementing granular consent procedures that allow users to specify specific goals for data gathering can enhance transparency and control.

3. Anonymization and Minimization: Minimizing the acquisition of personal data and adopting techniques like anonymization might lessen privacy issues.

The Imperative of Informed Consent

1. User Empowerment: Informed consent allows individuals to make decisions about their personal data, providing a sense of control and agency.

2. Trust: Consent promotes trust between users and businesses, boosting relationships and enabling ethical data activities.

3. Ethical Responsibility: Organizations have an ethical obligation to guarantee that users are fully aware of how their data will be used and to seek explicit consent.

Striking the Right Balance

1. Innovation and Progress: Data-driven innovation can lead to amazing improvements in technology, healthcare, and other industries, helping society as a whole.

2. Individual Rights: Protecting individual rights to privacy and data control is crucial for maintaining personal dignity and autonomy.

3. Collective Impact: Balancing data privacy, consent, and innovation has far-reaching effects that influence the societal impact of technology.

Conclusion

The delicate interplay between data privacy, permission, and innovation underlines the complex character of the digital world. While innovation pulls us toward new frontiers, maintaining individuals' rights to privacy and data control is equally crucial. Striking the appropriate balance entails not just following to legislation but also embracing ethical principles that prioritize transparency, fairness, and user empowerment. As technology continues to evolve, the ethical obligation to maintain data privacy and get informed consent remains crucial. By respecting individuals' rights and developing a culture of responsible data practices, we can negotiate the delicate balancing between innovation and maintaining individual dignity in a data-driven future.

Chapter 6

Fairness and Bias Mitigation

6.1   Quantifying Bias in AI Models: Unveiling Hidden Prejudices

Artificial intelligence (AI) models have the potential to change different industries, making choices faster and more efficiently. However, they are not immune to the biases that pervade society, and these biases can lead to unjust or discriminating outcomes. Quantifying bias in AI models has become a vital activity in the goal of ethical and responsible AI research. This chapter looks into the necessity of quantifying bias, covering the problems, methodologies, ramifications, and the broader impact of unveiling latent prejudices within AI systems.

Understanding Bias in AI Models

Bias in AI models refers to the presence of systematic and unjustified favouring or prejudice against specific groups or features. This bias might originate from biased training data, imbalanced representation, or subtle aspects in the data that influence model predictions in ways that may not correspond with fairness and equity.

The Significance of Quantifying Bias

Quantifying bias is a vital step in understanding and addressing bias in AI algorithms. It allows developers, researchers, and policymakers to quantify the level of bias, identify its origins, and take specific efforts to limit its impact. Without measurement, bias might remain hidden, leading to unintended effects and perpetuating unfair imbalances.

Challenges in Quantifying Bias

1. Complexity of Bias: Bias can manifest in numerous forms, making it tough to construct a single statistic to capture all sorts of bias.

2. Data Limitations: Accurate quantification requires access to diverse and representative data, which might be sparse or unavailable.

3. Context Sensitivity: Determining what constitutes bias can be context-sensitive, as fairness principles might vary across different areas and applications.

Methods for Quantifying Bias

1. Disparate Impact Analysis: This method examines the outcomes for different demographic groups and assesses whether there are inequalities that imply bias.

2. Statistical Parity: Statistical methods are employed to ensure that the chance of a favorable outcome is approximately equal across diverse groups.

3. Fairness Metrics: Metrics including equal opportunity, demographic parity, and disparate mistreatment quantify multiple elements of bias.

Ethical Implications of Quantifying Bias

1. Fairness: Quantifying bias coincides with the ethical concept of fairness, ensuring that AI systems do not unfairly disadvantage specific populations.

2. Accountability: Organizations and developers are accountable for the impact of their AI systems. Quantifying prejudice helps hold them responsible for any unfair outcomes.

3. Transparency: Transparently assessing bias strengthens the trust consumers place in AI systems, knowing that developers are actively addressing any biases.

The Broader Impact of Quantifying Bias

1. reduction: Quantification of bias is the first step toward successful bias reduction, leading to AI systems that are more egalitarian and accountable.

2. Inclusivity: By quantifying and resolving bias, AI systems become more inclusive and better able to serve diverse user populations.

3. Societal Trust: Unveiling latent prejudices through quantification improves trust between AI developers and consumers, encouraging good societal impact.

Addressing Bias Beyond Quantification

1. Algorithmic Design: Bias reduction should be included into the whole AI development lifecycle, from data gathering to model training and evaluation.

2. varied Representation: Ensuring varied representation in training data and development teams can help lessen the danger of bias.

3. Continuous Monitoring: Regularly monitoring AI systems for bias might assist resolve growing bias patterns and adapt to changing situations.

Conclusion

Quantifying bias in AI algorithms is not only a technological difficulty but also a moral obligation. As AI systems increasingly influence human decisions and interactions, it is crucial to guarantee that these systems do not perpetuate current biases or develop new kinds of discrimination. Quantification enables us to cast a light on latent prejudices, encouraging a more inclusive, equitable, and responsible AI landscape. By quantifying bias, we enable ourselves to address the ethical components of AI development, making informed decisions that uphold fairness, transparency, and social effect. In our quest for AI that fits with human values, quantifying bias appears as a vital step toward achieving trustworthy and ethical AI systems.

6.2 Fairness Metrics and Approaches: Navigating Equity in AI Systems

The advancement of artificial intelligence (AI) has provided transformative possibilities across different areas, but it has also underlined the significance of fairness and equity in technology. Fairness measurements and techniques have developed as crucial tools to ensure that AI systems make impartial decisions and do not discriminate against particular populations. This chapter addresses the significance of fairness measures, the difficulties they solve, various techniques to assessing fairness, and their role in developing a more just and equitable AI landscape.

Understanding Fairness Metrics

Fairness metrics are quantitative indicators used to analyse whether AI models and systems are making judgements without favoring or discriminating against specific demographic groups. These metrics give a systematic technique to discover and quantify bias or discrepancies in predictions, ensuring that AI systems respect ethical and egalitarian ideals.

Challenges Addressed by Fairness Metrics

1. Bias Detection: Fairness measures assist detect bias in AI systems by quantifying the disparities in outcomes across demographic groups.

2. Transparency: Fairness metrics help to the transparency of AI systems by giving objective measurements of fairness that can be reported to stakeholders.

3. Accountability: These measures hold developers and organizations accountable for the impact of their AI systems, generating a sense of responsibility.

Approaches to Measuring Fairness

1. Disparate Impact Analysis: This approach examines whether the outputs of an AI system disproportionately benefit or disfavor particular groups based on protected qualities.

2. Equal Opportunity: This statistic assesses if the genuine positive rates (recall) for different groups are similar, guaranteeing that pleasant outcomes are equally likely for all groups.

3. Demographic Parity: Also known as statistical parity, this strategy seeks to attain identical predicted odds for all groups, regardless of protected qualities.

4. Disparate Mistreatment: This statistic focuses on false positive and false negative rates, ensuring that errors are equally distributed among different groups.

Challenges in Fairness Metrics

1. Defining Fairness: Different fairness measurements could contradict with each other, making it challenging to describe a globally fair AI system.

2. Trade-offs: Striving for justice can occasionally result in trade-offs with other vital goals, such as accuracy and utility.

3. environment Sensitivity: What is considered fair can vary depending on the environment and the domain in which the AI system is being deployed.

Ethical Implications of Fairness Metrics

1. Fair Treatment: Fairness metrics align with the ethical notion of fair treatment, guaranteeing that persons are not unfairly favored or disadvantaged by AI systems.

2. Inclusivity: These measures contribute to the inclusivity of AI systems by encouraging equitable outcomes for varied user groups.

3. User Trust: Incorporating fairness metrics promotes user trust by demonstrating a commitment to openness, accountability, and ethical AI practices.

The Broader Impact of Fairness Metrics

1. Equity: Fairness metrics contribute to a more fair society by minimising the continuation of biases and disparities in AI-driven judgements.

2. Social Responsibility: Implementing fairness metrics shows an organization's social responsibility to design AI systems that match with societal ideals.

3. Policy and Regulation: Fairness metrics play a role in defining AI-related rules and regulations by giving objective indicators of bias and prejudice.

Balancing Fairness and Accuracy

1. Trade-offs: Striking the correct balance between fairness and accuracy involves careful analysis of the context, application, and user needs.

2. Algorithmic Design: Fairness should be built into the design of AI algorithms from the outset, rather than being handled as an afterthought.

3. Ongoing Monitoring: Regularly monitoring AI systems for fairness guarantees that developing bias patterns be discovered and remedied swiftly.

Conclusion

Fairness measurements and techniques are key instruments in the pursuit of ethical and equitable AI systems. As technology gradually enters every part of our lives, ensuring that AI choices are unbiased and fair becomes vital. These metrics give a formal means to assess inequities and bias, enabling developers, academics, and policymakers to take corrective actions. By incorporating fairness issues into the heart of AI development, we not only produce more accountable and transparent systems but also contribute to a more just and inclusive society. Fairness measures serve as a compass leading AI development toward a future where technology progress is harmonized with ethical ideals, equity, and societal effect.

6.3   Strategies for Mitigating Bias in Deep Learning: Navigating Ethical and Technical Challenges

Deep learning has taken artificial intelligence (AI) to unparalleled heights, performing astounding accomplishments in image identification, natural language processing, and more. However, these advances are not immune to biases present in training data, which might lead to biased and prejudiced AI systems. Strategies for minimising bias in deep learning have become crucial to assure ethical and responsible AI research. This chapter analyses the significance of these tactics, the difficulties they address, methods of implementation, and their role in constructing AI systems that promote fairness, equity, and inclusivity.

Understanding Bias in Deep Learning

Bias in deep learning refers to the presence of systematic favouring or prejudice toward specific groups or traits in AI algorithms. prejudiced training data, historical inequities, and societal prejudices can all contribute to the formation of prejudiced AI systems.

The Importance of Mitigating Bias

Mitigating bias in deep learning is critical for various reasons:

1. Equity: Bias prevention aids to establishing AI systems that make fair and equitable decisions for all persons and groups.

2. Ethical obligation: Developers have an ethical obligation to ensure that AI systems do not perpetuate or magnify existing biases.

3. Trust: Mitigating prejudice promotes user trust in AI systems, enabling wider acceptance and deployment.

Challenges in Mitigating Bias in Deep Learning

1. Hidden Bias: Biases can be subtle and difficult to identify, making them challenging to control successfully.

2. impair on Performance: Addressing bias can occasionally impair model performance, leading to trade-offs between accuracy and fairness.

3. Complex Bias Patterns: Some biases show in complex patterns that are not easily captured by typical mitigation strategies.

Strategies for Mitigating Bias in Deep Learning

1. varied and Representative Training Data: Using varied and representative training data might help avoid bias from being amplified in AI models.

2. Data Augmentation: Augmenting data through transformations can balance the representation of different groups, decreasing bias.

3. Preprocessing Techniques: Preprocessing techniques like re-sampling and re-weighting help decrease bias by changing the representation of distinct groups.

4. Fairness-Aware Algorithms: Developing algorithms that explicitly account for fairness helps prevent biased outcomes during training.

5. Post-hoc Analysis: Analyzing model predictions post-training can help discover and fix biased trends.

Ethical Implications of Bias Mitigation

1. Fairness: Mitigating bias aligns with the ethical value of fairness, ensuring that AI systems treat all individuals and groups fairly.

2. Inclusivity: Bias mitigation increases inclusivity by designing AI systems that do not discriminate against any population.

3. Accountability: Organizations and developers are accountable for the impact of their AI systems. Bias mitigation reflects appropriate AI development.

The Broader Impact of Bias Mitigation

1. Social Equity: Bias mitigation leads to a more just and equitable society by preventing AI systems from perpetuating detrimental imbalances.

2. User Trust: Users are more inclined to trust AI systems that have been created to be unbiased and fair.

3. Regulation and Policy: Bias mitigation solutions influence the development of legislation and policies relating to AI ethics and fairness.

Balancing Bias Mitigation and Performance

1. Trade-offs: Striking the correct balance between bias reduction and performance involves careful analysis of the application and user needs.

2. Algorithmic Design: Bias reduction should be an essential element of the algorithm's design, rather than an afterthought.

3. Continuous Monitoring: Regularly monitoring AI systems for bias guarantees that emergent bias patterns are discovered and rectified swiftly.

Conclusion

Strategies for minimising bias in deep learning are vital for developing ethical, accountable, and equitable AI systems. As AI grows to integrate into various facets of our life, tackling bias becomes a moral obligation. These tactics not only lead to more responsible AI development but also establish systems that uphold societal values of fairness and diversity. By accepting these tactics, we embark on a road toward AI that respects human dignity, minimizes societal harm, and promotes good society impact. Mitigating bias in deep learning is a cornerstone in designing AI systems that promote ethics, transparency, and social equity.

Chapter 7
Case Studies in Ethical and Explainable AI

7.1   Healthcare: Diagnostics, Treatment, and Patient Privacy

The convergence of healthcare and technology has led to significant improvements in diagnoses, treatment, and patient care. From precision medicine to telehealth, breakthrough technologies are revolutionising the way medical professionals diagnose ailments, offer treatments, and communicate with patients. However, these technical developments also present difficult ethical and privacy problems. This chapter addresses the expanding environment of healthcare, the role of technology in diagnostics and treatment, and the requirement to combine innovation with patient privacy.

Technology's Role in Healthcare Diagnostics and Treatment

1. Precision Medicine: Advances in genetics and data analytics enable individualised treatment strategies based on an individual's genetic composition, enhancing outcomes and reducing unwanted effects.

2. Medical Imaging: Technologies like MRI, CT scans, and ultrasound provide detailed images that aid in accurate diagnosis and treatment planning.

3. AI-Assisted Diagnostics: Artificial intelligence algorithms evaluate medical pictures and data to boost diagnostic accuracy and speed, aiding physicians in making educated decisions.

4. Telehealth and Remote Monitoring: Telehealth platforms allow patients to consult with healthcare professionals remotely, boosting access to care and monitoring chronic illnesses from home.

5. Robot-Assisted Surgery: Robotic technologies enable surgeons in conducting difficult procedures with precision, decreasing invasiveness and patient recovery time.

Benefits of Technology in Healthcare

1. Early Detection: Advanced diagnostics enable early detection of diseases, boosting the likelihood of successful treatment.

2. Customized Treatment: Personalized medicine tailors treatment plans to specific patients, maximising therapeutic outcomes.

3. Improved Efficiency: Technology improves administrative operations, decreasing paperwork and allowing medical staff to focus more on patient care.

4. rural Access: Telehealth promotes healthcare access for rural or underserved locations, bridging geographical constraints.

Patient Privacy and Data Security

1. Electronic Health Records (EHRs): Digitized health records enhance efficient data interchange among healthcare providers but also raise worries about data breaches and illegal access.

2. Data Sharing: The sharing of patient data between institutions can lead to greater care coordination but also demands rigorous data protection measures.

3. Cybersecurity: The digitalization of healthcare systems raises the vulnerability to hackers, placing sensitive patient data at risk.

Ethical Considerations in Healthcare Technology

1. Informed Consent: Patients must be fully informed about the collection, use, and sharing of their health data while utilising digital health tools.

2. Data Ownership: Clarifying who owns and controls patient health data is vital, ensuring individuals have agency over their information.

3. Equity: Ensuring fair access to healthcare technologies is crucial to prevent increasing current healthcare inequities.

4. Transparency: Healthcare institutions must be upfront about how patient data is used and ensure that data usage matches with patient expectations.

Challenges in Balancing Innovation and Patient Privacy

1. Data Accessibility: Striking a balance between data accessibility for research and treatment reasons while respecting patient privacy is tough.

2. Consent Complexity: Obtaining meaningful and informed patient consent for data sharing can be complicated, given the technical and legal complexity.

3. legislative Frameworks: Rapid technological improvements outstrip legislative frameworks, providing a difficulty to provide effective patient privacy protections.

The Imperative of Patient Privacy

1. Trust: Maintaining patient privacy creates trust between patients and healthcare professionals, encouraging open communication and care-seeking behaviors.

2. Legal and Ethical Obligations: Healthcare practitioners and institutions have a legal and ethical obligation to secure patient data.

3. Patient Autonomy: Respecting patient privacy supports their autonomy and control over their own health information.

Conclusion

The incorporation of technology into healthcare has transformed diagnoses, treatment, and patient care, boosting efficiency and outcomes. However, as technology progresses, the ethical and privacy issues of patient data gathering, sharing, and utilisation become paramount. Striking the delicate balance between innovation and patient privacy is crucial to guarantee that healthcare technology preserve ethical norms, respect patient autonomy, and sustain confidence. As the healthcare landscape continues to advance, integrating patient privacy alongside technological progress is crucial to building a healthcare system that is not just efficient and effective but also ethically sound and respectful of patient rights.

7.2 Finance: Algorithmic Trading and Responsible Lending

The banking business has seen a major upheaval with the arrival of technology, changing how financial institutions trade and communicate with clients. Algorithmic trading has changed the speed and efficiency of executing deals, while responsible lending procedures attempt to ensure that financial institutions deliver loans in a fair and ethical manner. This chapter discusses the dynamic environment of finance, the importance of algorithmic trading, the imperative of responsible lending, and the problems of balancing innovation with ethical considerations.

Algorithmic Trading: Enhancing Efficiency and Complexity

1. Definition and Mechanism: Algorithmic trading involves the use of pre-programmed algorithms to execute trades at high speeds and volumes, exploiting market inefficiencies and opportunities.

2. Speed and Precision: Algorithms process data and execute trades in milliseconds, limiting human interaction and lowering execution errors.

3. Market Liquidity: Algorithmic trading promotes market liquidity by providing constant buying and selling opportunities.

4. Complexity: The complexity of algorithms raises worries about market stability, as quick trading can lead to unpredictable price movements.

Benefits and Challenges of Algorithmic Trading

1. Efficiency: Algorithmic trading promotes market efficiency by lowering bid-ask spreads and ensuring trades are performed at appropriate pricing.

2. Risk Management: Algorithms can automatically apply risk management methods, including as stop-loss orders, to limit prospective losses.

3. Regulatory Concerns: The speed and intricacy of algorithmic trading have triggered regulatory scrutiny to prevent market manipulation and assure fairness.

4. Ethical Implications: The employment of algorithms poses ethical considerations linked to transparency, accountability, and potential biases in trading decisions.

Responsible Lending: Ethical Financial Services

1. Definition and Principles: Responsible lending entails delivering loans to consumers in a transparent, fair, and ethical manner that aligns with their financial capabilities.

2. Customer-Centric Approach: Financial institutions must analyse borrowers' ability to repay loans and provide clear terms and conditions.

3. Risk Mitigation: Responsible lending procedures minimise overindebtedness, minimising the risk of financial instability for borrowers and lenders.

The Importance of Responsible Lending

1. Consumer Protection: Responsible lending protects consumers from exploitative activities and ensures they can make informed financial decisions.

2. Financial Stability: By limiting excessive borrowing, responsible lending contributes to the stability of both individual households and the financial system.

3. Trust: Adhering to responsible lending norms creates trust between financial institutions and borrowers, encouraging long-term relationships.

Challenges in Balancing Innovation and Responsibility

1. Algorithmic Trading Complexity: The sophisticated nature of algorithms makes it challenging to predict their behavior, leading to concerns about market stability.

2. Fairness and Bias: Algorithmic trading systems might unwittingly perpetuate biases contained in previous data, leading to unjust trading outcomes.

3. Responsible Lending Dynamics: Balancing the desire of profit with the ethical obligation of responsible lending can create difficulties for financial organisations.

Ethical Considerations in Finance

1. Transparency: Transparent methods in algorithmic trading and lending guarantee that clients understand the processes and dangers involved.

2. Equity: Both algorithmic trading and lending should be designed to prevent benefitting particular persons or groups unfairly.

3. Accountability: Financial organisations are accountable for the effects of their trading decisions and lending activities.

The Broader Impact of Ethical Finance Practices

1. Market Integrity: Ethical behaviours protect the integrity of financial markets, promoting fair competition and investor trust.

2. Economic Stability: Ethical finance practices assist to general economic stability by preventing financial crises caused by reckless lending.

Conclusion

Algorithmic trading and responsible lending reflect two elements of the developing finance world, where technology innovation meets ethical responsibility. While algorithmic trading promotes efficiency and liquidity, it requires strict control to minimise unforeseen outcomes. Responsible lending guarantees that financial services are offered honestly and ethically, protecting clients from financial difficulty. Balancing innovation with ethical considerations is a problem that demands vigilance, regulatory assistance, and a commitment to safeguard market integrity, consumer protection, and financial stability. As technology continues to transform the banking industry, it is vital to welcome innovation while respecting ethical values to create a robust and equitable financial ecosystem.

7.3 Criminal Justice: Predictive Policing and Bias Detection

The combination of technology and criminal justice has led to both potential breakthroughs and ethical issues. Predictive policing strives to prevent crime by employing data analytics and machine learning, whereas bias detection seeks to find and fix prejudices in the criminal justice system. This chapter dives into the intricacies of predictive policing, the significance of bias detection, the ethical problems they bring, and the necessity to strike a balance between innovation and justice.

Predictive Policing: Transforming Law Enforcement

1. Definition and Function: Predictive policing utilizes data analysis and machine learning to estimate where and when crimes are likely to occur, allowing law enforcement organisations to spend resources more effectively.

2. statistics Sources: Predictive policing relies on past crime statistics, socioeconomic factors, and other pertinent information to develop forecasts.

3. Crime Prevention: The purpose of predictive policing is to prevent crimes before they happen by identifying regions with higher potential for criminal activity.

4. Challenges: Predictive policing raises questions regarding privacy, accuracy, accountability, and potential biases in algorithmic predictions.

Benefits and Concerns of Predictive Policing

1. Crime Prevention: Predictive policing has the ability to lower crime rates and improve public safety by focusing law enforcement resources more strategically.

2. Resource Allocation: By focusing resources on high-risk locations, police departments may improve their response and dissuade criminal activity.

3. Bias and Discrimination: If not appropriately deployed, predictive police algorithms might perpetuate existing biases and unfairly target certain communities.

Bias Detection: Uncovering Systemic Prejudices

1. Bias in Criminal Justice: Bias can appear in numerous stages of the criminal justice process, from arrest to sentencing, hurting vulnerable communities disproportionately.

2. Data-Driven Discrimination: Biases existing in previous crime data can lead to biased algorithmic predictions and judgements, perpetuating inequities.

3. Ethical Imperative: Detecting and eliminating biases is vital for developing a more just and equitable criminal justice system.

The Significance of Bias Detection

1. Fairness: Bias detection supports fairness in the criminal justice system by identifying and rectifying unfair acts.

2. Transparency: Uncovering biases enhances transparency, allowing stakeholders to identify and address systemic concerns.

3. Accountability: Bias detection makes criminal justice authorities accountable for their actions and judgements, promoting more trust from the public.

Challenges in Balancing Innovation and Justice

1. Algorithmic intricacy: The intricacy of predictive policing algorithms might make it tough to grasp how decisions are made.

2. Data Quality: Biased past crime data can lead to biased algorithmic forecasts, reinforcing existing imbalances.

3. Responsible Implementation: Implementing predictive policing and bias detection technology demands careful evaluation of their potential repercussions.

Ethical Considerations in Criminal Justice

1. Equity: Ensuring that algorithms do not disproportionately target specific communities is vital for an equitable criminal justice system.

2. Transparency: Law enforcement agencies must be upfront about their use of predictive policing algorithms and their efforts to remove biases.

3. Community Engagement: Involving communities in the development and deployment of predictive policing technologies can help address concerns and enhance outcomes.

The Broader Impact of Ethical Criminal Justice Practices

1. minimising Inequities: Ethical practices in predictive policing and bias detection contribute to minimising inequities in the criminal justice system.

2. Trust and Confidence: Ethical practices increase public trust and confidence in law enforcement institutions, enabling collaboration and community support.

Conclusion

Predictive policing and prejudice detection lie at the confluence of technological innovation and ethical fairness. While predictive policing holds the potential to boost crime prevention and resource allocation, its implementation requires safeguards against biases that could perpetuate inequity. Bias identification, on the other hand, is crucial for exposing systemic prejudices and rectifying them to create a more equal criminal justice system. Striking a balance between innovation and justice involves a commitment to transparency, accountability, and community participation. As technology continues to evolve, it is crucial to guarantee that advancements in criminal justice are driven by ethical considerations, leading to a more equitable, transparent, and accountable system that preserves the rights and dignity of all individuals.

Chapter 8

Regulations and Guidelines

8.1 Overview of Current AI Regulations: Navigating the Evolving Landscape

As artificial intelligence (AI) technology continues to evolve at an unprecedented speed, governments and regulatory organisations worldwide are dealing with the need to build frameworks that balance innovation with ethical considerations, safety, and accountability. The current landscape of AI rules is characterized by a combination of national and international activities, reflecting the multifaceted difficulties posed by AI's revolutionary potential. This chapter presents an overview of the current state of AI legislation, addressing important areas of attention, problems, and the requirement to develop a harmonious and responsible regulatory system.

The Need for AI Regulations

1. Ethical Concerns: As AI systems become more autonomous and omnipresent, ethical concerns such as bias, justice, privacy, and responsibility have risen to the forefront.

2. Safety: Ensuring the safety of AI systems, especially in important fields like autonomous vehicles and healthcare, is a paramount problem.

3. Transparency: The lack of transparency in AI decision-making processes raises issues about the understandability and fairness of AI-driven outcomes.

4. Market Fairness: Regulations are essential to prevent market monopolization and foster fair competition among AI businesses.

International Efforts in AI Regulation

1. European Union: The EU has taken a leading position in AI governance with the General Data Protection governance (GDPR) and the proposed Artificial Intelligence Act, emphasizing transparency, accountability, and human oversight.

2. United States: The U.S. lacks comprehensive government AI legislation but has focused on recommendations for AI ethics, safety, and funding for AI research and development.

3. International Collaboration: Organizations like the United Nations and the Organisation for Economic Co-operation and Development (OECD) are working on rules and principles for ethical AI.

Key Areas of Focus in AI Regulations

1. Data Privacy and Protection: Regulations like the GDPR and California Consumer Privacy Act (CCPA) underscore the necessity for transparent data collection, storage, and usage policies in AI systems.

2. Bias and justice: Regulations strive to prevent algorithmic bias and prejudice, supporting justice and equity in AI-driven choices.

3. Safety and Liability: Regulations in autonomous systems, such as self-driving cars, address safety standards, testing, and liability in case of accidents.

4. Accountability and openness: Regulations compel AI developers to disclose reasons for automated judgements, ensuring accountability and openness.

Challenges in AI Regulations

1. Rapid Technological Advancements: AI technology grows faster than legislation can be created and enforced, resulting to potential gaps.

2. Lack of Universal Definitions: Defining AI and its scope varies across rules and jurisdictions, generating ambiguity.

3. Global Consistency: Ensuring similar and harmonized legislation across different countries is a difficulty due to varied cultural, legal, and economic circumstances.

The Imperative of Responsible Regulation

1. Ethical AI Development: Regulations play a significant role in supporting the development of AI systems that comply with ethical principles and social values.

2. Trust and Adoption: Responsible rules promote public trust and confidence in AI technologies, enabling wider adoption and acceptance.

3. Accountability and Liability: Regulations hold AI developers and organizations accountable for the impact of their systems, ensuring they take adequate measures.

Balancing Innovation and Regulation

1. Innovation Stifling: Overly rigid laws can stifle innovation, hindering the development of cutting-edge AI technologies.

2. Ethical Considerations: Striking the correct balance between innovation and regulation demands emphasising ethical considerations while still allowing for technological progress.

3. Flexibility: Regulations should be adaptive to accommodate the developing nature of AI technology while retaining ethical safeguards.

Conclusion

The current landscape of AI rules reflects the complex issues created by the rapid breakthroughs of AI technology. Balancing innovation with ethical, safety, and accountability considerations is at the heart of crafting effective regulations. As governments and international organizations work toward building comprehensive frameworks, it is crucial to facilitate collaboration and harmonization to guarantee that

AI rules address the complex concerns while supporting responsible AI research. A wise and responsive regulatory environment will play a critical role in shaping the future of AI and its influence on society.

8.2 Ethical Frameworks and Guidelines for AI Development: Charting the Path to Responsible Innovation

As artificial intelligence (AI) technologies become increasingly integrated into all facets of our life, the necessity for ethical considerations in their development and deployment has assumed paramount importance. Ethical frameworks and standards give an organised method to ensuring that AI systems are developed, implemented, and used in ways that correspond with societal values, respect human rights, and avoid potential harms. This chapter looks into the significance of ethical frameworks in AI development, analyses essential principles and rules, and emphasizes the role they play in supporting responsible innovation.

The Need for Ethical Frameworks in AI

1. Complex Ethical Challenges: AI systems can effect privacy, bias, discrimination, accountability, and more, needing strong ethical rules.

2. Human-Centered Approach: Ethical frameworks ensure that AI technologies are designed to benefit humanity, without compromising on fundamental rights.

3. Trust and Adoption: Ethical AI development increases public trust, which is crucial for widespread adoption and acceptance of AI technologies.

Key Principles of Ethical Frameworks

1. Transparency: AI systems should be designed and deployed transparently, ensuring that their operation is intelligible and explainable.

2. Fairness and Equity: AI technology must be created with fairness in mind, minimising biases and supporting equal treatment of all individuals.

3. Privacy: Respect for user privacy is vital, necessitating stringent data protection procedures and informed permission.

4. Accountability: Developers and organizations must be held accountable for the repercussions of AI systems, ensuring they take responsibility for their activities.

5. Beneficence: AI technology should try to increase benefits while limiting potential damages to individuals and society.

Global Ethical AI Guidelines

1. IEEE Global Initiative on Ethics of Autonomous and Intelligent Systems: Provides a comprehensive set of rules emphasising on openness, accountability, and data agency.

2. European Commission's Ethics Guidelines for Trustworthy AI: Emphasizes human agency and oversight, technical robustness, and societal well-being.

3. UNESCO's Recommendation on AI Ethics: Advocates for diversity, sustainability, and human dignity in AI development.

Ethical AI in Specific Domains

1. Healthcare: Ethical AI principles in healthcare emphasise on patient autonomy, privacy, and unbiased treatment suggestions.

2. Finance: Ethical AI in finance emphasises justice, transparency, and accountability in algorithmic trading and lending.

3. Criminal Justice: Ethical AI paradigms emphasize fairness, accountability, and bias detection in predictive policing and decision-making.

Challenges in Ethical AI Development

1. Bias and Fairness: Addressing biases in AI systems and guaranteeing fair outcomes presents technological and ethical problems.

2. Algorithmic Complexity: The complexity of AI algorithms makes it challenging to grasp their decision-making processes.

3. Cultural Variability: Ethical considerations might vary between cultures, requiring flexible frameworks that acknowledge cultural variation.

Implementing Ethical Frameworks

1. Cross-Disciplinary Collaboration: Collaboration between AI researchers, ethicists, legal experts, and other stakeholders is vital to establishing complete frameworks.

2. Human-Centric Design: AI systems should prioritize human well-being and align with human values in their design and functionality.

3. Ongoing Assessment: Ethical frameworks should be regularly assessed and revised to keep pace with new AI technologies and emerging concerns.

Balancing Innovation and Ethics

1. Innovation-Driven Ethical Considerations: Ethical frameworks should foster innovation while ensuring that emerging AI technologies adhere to ethical ideals.

2. Responsible Deployment: Developers and organizations must examine the ethical implications of AI throughout its lifecycle, from development to deployment.

Conclusion

Ethical frameworks and guidelines play a crucial role in directing the development and deployment of AI systems that are consistent with societal values, human rights, and ethical issues. As AI continues to alter businesses and influence daily life, it is crucial to prioritize responsible innovation. By adhering to ethical values such as transparency, justice, accountability, and beneficence, the AI community may create a future where technology acts as a force for positive societal effect, respects individual rights, and contributes to the advancement of mankind as a whole.

8.3 Collaborative Efforts: Industry and Government Roles in Shaping the Future

In the fast expanding environment of technology and innovation, collaborative activities between industry and government play a critical role in defining the future of numerous sectors, from technological creation to policy formulation. This dynamic cooperation brings together the expertise, resources, and regulatory oversight needed to promote innovation while guaranteeing ethical, safe, and responsible advancement. This chapter addresses the relevance of joint efforts, the roles that industry and government play, the obstacles they encounter, and the importance of balancing interests for the greater good.

The Power of Collaboration

1. Innovation Acceleration: Collaborations between industry and government accelerate innovation by combining private sector experience and government resources.

2. Regulatory Alignment: Collaborative efforts assist ensure that rules keep pace with technology changes, maintaining a balanced regulatory environment.

3. Resource Synergy: Industry brings finance, research, and development capabilities, while government contributes funding, regulatory control, and societal interests.

Roles of Industry

1. Innovation Drivers: Industries are at the forefront of technological breakthroughs, driving innovation and research in numerous domains.

2. Resource Allocation: Industries allocate resources for research, development, and commercialization, contributing to technological progress.

3. Expertise: Industry professionals offer insights into emerging trends, possible applications, and practical concerns for deploying technology.

Roles of Government

1. Regulatory Oversight: Governments offer regulatory frameworks that protect the safety, ethics, and ethical use of technologies.

2. Public Interest: Government's responsibility involves defending public interests, privacy, and fair competition, even as industries pursue innovation.

3. financing and Support: Governments typically give financing, grants, and support for research efforts that match with national interests and priorities.

Industry and Government Collaboration: Examples

1. Healthcare and Biotechnology: Collaborations between pharmaceutical corporations and government organisations

contribute to quicker medication development while addressing safety and ethical issues.

2. Environmental Conservation: Partnerships between clean energy companies and government bodies facilitate the transition to renewable energy sources while addressing environmental issues.

3. Technology and Standards: Industry-government collaboration produces technical standards that encourage interoperability and stimulate innovation across industries.

Challenges in Collaborative Efforts

1. Conflicting Interests: Industry and government can have conflicting interests, such as profit incentive versus public safety, necessitating cautious negotiation.

2. Regulatory Lag: Rapid technical improvements can exceed regulatory frameworks, causing issues in guaranteeing ethical and safe adoption.

3. Transparency and Accountability: Striking the correct balance between industry's demand for proprietary information and government's requirement for transparency can be tough.

Balancing Interests for the Greater Good

1. Innovation and Ethical factors: Industry and government must collaborate to ensure that innovation corresponds with ethical, societal, and environmental factors.

2. Responsibility and Liability: Collaboration guarantees that industry are responsible for the repercussions of their discoveries, while governments assure accountability.

3. Public Trust: Collaborative efforts that highlight ethical issues contribute to public trust in both technology and governance.

The Imperative of Public Involvement

1. Stakeholder Engagement: Including public and civil society voices in collaborative activities ensures that varied perspectives are acknowledged.

2. Democratic Decision-Making: Collaboration that involves public involvement helps ensure that technology development aligns with democratic norms.

Conclusion

Collaborative efforts between industry and government serve as the cornerstone of responsible innovation and technological growth. While industries drive innovation, government offers oversight, regulation, and responsibility to ensure that technology aligns with social values and public interests. Striking the correct balance between innovation and ethical issues involves honest communication, shared aims, and a dedication to the greater good.

As technology continues to alter our world, creative relationships between industry and government will be necessary to design a future that is both inventive and responsible, benefitting individuals, society, and the planet.

Chapter 9
Charting a Responsible AI Future

9.1   Designing AI for Positive Societal Impact: Ethical Responsibility and Innovation

Artificial intelligence (AI) has the ability to change businesses, enhance quality of life, and solve challenging challenges. However, this promise comes with ethical duties to ensure that AI technologies are built and deployed for positive society effect. This chapter digs into the need of developing AI with ethical considerations, analyses fundamental concepts for creating good AI systems, and emphasizes the role of creativity in accomplishing permanent positive change.

Ethical Imperatives in AI Design

1. Preventing Harm: Ethical AI design focuses avoiding harm to individuals, groups, and society at large, ensuring that AI systems do not perpetuate prejudices, discrimination, or privacy breaches.

2. Maximizing Benefit: AI design should attempt to enhance advantages for individuals and society while limiting potential negative outcomes.

3. Transparency: Ethical AI design stresses transparency, ensuring that people understand how AI systems make decisions and operate.

Key Principles for Designing AI for Positive Impact

1. Human-Centric Approach: AI systems should prioritize human well-being, boosting human capacities and augmenting human decision-making rather than replacing it.

2. Fairness and Inclusivity: Designing AI systems that are fair, unbiased, and inclusive, guaranteeing that all persons, regardless of demographics, receive equitable treatment.

3. Privacy and Data Protection: Protecting user privacy through responsible data collecting, storage, and usage methods is important to ethical AI design.

4. Accountability and Explainability: AI systems should be accountable for their actions, and their decision-making processes should be explainable to users.

5. society Impact study: AI development should include rigorous study of potential society implications, both good and negative, to support responsible decision-making.

Innovation for Positive Societal Impact

1. Problem Solving: AI innovation should be driven by solving real-world problems, tackling concerns that have a direct impact on individuals, communities, and the environment.

2. Social and Environmental Goals: Innovators should integrate AI breakthroughs with social and environmental goals, addressing concerns such as climate change, public health, and education.

3. Community Involvement: Involving varied populations in the AI design process guarantees that technologies answer their particular requirements and concerns.

Examples of Positive Societal Impact using AI

1. Healthcare: AI-powered diagnostics and treatment recommendations improve patient outcomes and enable personalized therapy.

2. Environmental Conservation: AI algorithms aid in data analysis for monitoring climate change, deforestation, and wildlife conservation.

3. Education: AI-driven personalized learning platforms enhance educational experiences, responding to individual learning styles and demands.

Challenges in Designing AI for Positive Impact

1. prejudice Mitigation: Addressing prejudice in AI systems needs careful data collecting, algorithmic design, and continual monitoring.

2. unanticipated repercussions: Ethical AI design tries to predict and minimise unanticipated negative repercussions that might result from AI deployment.

3. Balancing Interests: Balancing innovation with ethical issues necessitates reconciling the interests of multiple stakeholders, including developers, users, and society.

The Continuous Journey Towards Positive Impact

1. Iterative Design: Ethical AI design is an iterative process that requires constant assessment, adaptation, and improvement based on real-world feedback.

2. Collaborative Partnerships: Industry, academia, governments, and civil society must collaborate to ensure that AI breakthroughs contribute constructively to society.

Conclusion

Designing AI for beneficial societal impact is both an ethical responsibility and an opportunity for innovation. By adhering to ethical standards, emphasising human well-being, and aligning AI advancements with social and environmental goals, we can harness the transformative potential of AI to create a better world for present and future generations. Balancing innovation with ethical considerations ensures that technology serves humanity's best interests, supports ethical principles, and contributes to a more just, equitable, and sustainable future.

9.2   The Role of Education in Ethical and Explainable AI: Empowering Minds for Responsible Innovation

As artificial intelligence (AI) continues to transform industries and daily life, the need for education in ethical and explainable AI becomes increasingly critical. Education plays a vital role in preparing humans to understand, develop, and use AI technology in ways that comply with ethical principles and encourage transparency. This chapter discusses the role of education in encouraging responsible AI innovation, the important components of an effective educational system, and the larger societal impact of educated AI practitioners.

The Importance of Ethical and Explainable AI Education

1. Ethical Awareness: Education equips humans with the skills to understand ethical implications and dilemmas originating from AI technologies.

2. Responsible Development: Ethical and explainable AI education helps that developers construct systems that are transparent, fair, and accountable.

3. Informed Decision-Making: Education empowers users to make informed decisions about AI adoption and use, considering ethical and privacy considerations.

Components of an Effective AI Education Framework

1. basic grasp: Education starts with creating a basic grasp of AI principles, algorithms, and their real-world applications.

2. Ethical Considerations: Students learn about the ethical difficulties faced by AI, including bias, privacy, accountability, and potential societal repercussions.

3. Transparency and Explainability: Education highlights the need of designing AI systems that are transparent and explainable to users.

4. Case Studies: Analyzing real-world case studies helps learners understand the complexities of ethical AI concerns and decision-making.

5. Interdisciplinary Approach: AI education should integrate multiple fields, including computer science, ethics, law, and social sciences.

Benefits of Ethical and Explainable AI Education

1. Informed AI Practitioners: Education generates AI practitioners who are well-versed in ethical matters and devoted to responsible innovation.

2. Ethical AI Adoption: Educated persons are more inclined to adopt AI technology responsibly, considering their influence on society and individuals.

3. Preventing prejudice: Education empowers developers with tools to recognise and eliminate prejudice in AI systems, ensuring fair and equal outcomes.

Educational Initiatives in Ethical and Explainable AI

1. Academic Programs: Universities provide specific courses and degrees in AI ethics, responsible AI development, and explainable AI.

2. Online Resources: Open-access online platforms give instructional resources, tutorials, and case studies to a global audience.

3. Workshops and Seminars: Industry and academia collaborate to create workshops and seminars that promote ethical AI practices.

Challenges in AI Education

1. Rapid Technological Advancements: The growing nature of AI technology necessitates education to stay pace with the latest breakthroughs.

2. Interdisciplinary Complexity: Effective AI education needs collaboration between disciplines, which can be tough to achieve.

3. Global Consistency: Ensuring uniform AI education across diverse countries and cultures is a hard task.

Societal Impact of Ethical AI Education

1. Ethical Technologists: Education cultivates a generation of ethical AI technologists who promote responsible innovation.

2. Policy and Advocacy: Educated persons contribute to the creation of ethical AI policies and advocate for responsible AI use.

3. Public Trust: Ethically educated AI practitioners boost public trust in AI technologies, facilitating wider adoption and acceptance.

Empowering the Future of AI

1. Continuous Learning: AI education is a continual process that adjusts to technology breakthroughs and developing ethical concerns.

2. Collaborative Efforts: Governments, academia, industry, and civic society must work to build comprehensive and effective AI education initiatives.

Conclusion

The role of education in ethical and explainable AI is crucial to ensure that AI technologies are created and implemented in ways that match with society values and ethical issues. By training individuals with the knowledge and skills to manage the difficulties of AI ethics, transparency, and accountability, education empowers the next generation of AI practitioners to be responsible innovators. As AI continues to affect the future, a strong foundation in ethical and explainable AI education will contribute to a more just, egalitarian, and informed technological landscape.

9.3 Balancing Innovation and Accountability: Navigating the Ethical Landscape of Technological Progress

In the era of fast technological growth, the delicate balance between innovation and responsibility has become a major concern for industries, governments, and society at large. Innovation generates development, economic prosperity, and societal transformation, but it must be tempered with accountability to ensure that the benefits are broad and the potential risks are avoided. This chapter digs into the intricate interplay between innovation and accountability, analysing the relevance of establishing the appropriate equilibrium, the problems that arise, and the solutions to support responsible advancement.

The Dynamic of Innovation and Accountability

1. Innovation's Transformative Power: Innovation has the capacity to revolutionize sectors, create new markets, and solve complex global concerns.

2. Accountability's Imperative: While innovation moves society forward, accountability guarantees that its impacts are beneficial, equitable, and sustainable.

Significance of Finding Balance

1. Societal Impact: Balancing innovation and accountability eliminates unanticipated negative implications that can occur from unrestrained progress.

2. Public Trust: Responsible innovation builds public trust in industry and governments, enabling broader adoption of new technology.

3. Ethical Considerations: Striking a balance safeguards against the infringement of ethical norms, human rights, and privacy.

Challenges in Balancing Innovation and Accountability

1. Pace of Innovation: Rapid technical breakthroughs often outrun the creation of legal frameworks and ethical considerations.

2. unintended repercussions: Innovation can lead to unintended repercussions, such as biases, job displacement, and societal changes.

3. Stakeholder Interests: Different stakeholders, including industries, governments, and communities, can have divergent interests.

Strategies for Responsible Progress

1. Ethical by Design: Innovations should be built with ethical considerations at the core, mitigating potential harms and providing benefits.

2. Collaboration: Industries, governments, academia, and civil society must collaborate to promote responsible innovation and balanced accountability.

3. Regulatory Frameworks: Governments play a critical role in building regulatory frameworks that encourage innovation while protecting public interests.

4. Transparency and Accountability procedures: Ensuring transparency in innovation processes and developing procedures for accountability enhance responsible advancement.

Ethics in Technological Advancements

1. AI and Automation: As AI and automation alter sectors, they must be developed in ways that protect human dignity and job rights.

2. Emerging Technologies: Ethical considerations in areas like biotechnology, nanotechnology, and quantum computing are pivotal for responsible advancement.

The Role of Education and Awareness

1. Ethical Literacy: Educating individuals about the ethical implications of innovation empowers them to make informed decisions and demand responsibility.

2. Ethical Technology Professionals: Education generates a new generation of technologists who prioritize ethics and evaluate the societal impact of their discoveries.

Striking a Balance in Ethical Dilemmas

1. Privacy vs. Innovation: Innovations like data-driven technologies must find the balance between exploiting data and preserving individuals' privacy rights.

2. Safety vs. Speed: In domains like as autonomous vehicles, the trade-off between quick development and safety issues must be carefully controlled.

The Imperative of Continuous Adaptation

1. Evolving Ethical Landscape: The ethical implications of innovation always evolve, needing continuing adaptation and responsive regulation.

2. Ethics and AI: As AI systems become more autonomous, ethical questions regarding transparency, prejudice, and responsibility become crucial.

Conclusion

Balancing innovation and responsibility is a complicated, continuing process that involves the involvement of corporations, governments, academics, and civil society. Responsible innovation ensures that technical breakthroughs accord with society values, human rights, and ethical principles. Striking the correct balance needs forethought, interdisciplinary teamwork, solid regulatory frameworks, and an unshakable commitment to fostering progress while limiting harm. By negotiating this delicate equilibrium, society can harness the transformative force of innovation while maintaining the well-being and interests of all individuals and communities.

Conclusion

1. Reflections on the Journey: Navigating the Complex Path of Ethical AI

The route towards ethical artificial intelligence (AI) has been one of ongoing change, obstacles, and progress. As we reflect on this transformative journey, it becomes obvious that the quest of ethical AI is not only a destination, but a continuous process that involves awareness, collaboration, and a dedication to crafting a future that respects human values and enhances social well-being. This chapter digs into the essential insights obtained from the journey of ethical AI, the accomplishments accomplished, the difficulties faced, and the importance of continuous dedication to responsible AI development.

The Evolution of Ethical AI

1. Early problems: In the early days of AI, ethical problems were largely disregarded as the focus was on technical breakthroughs.

2. Ethics Rising to significance: As AI technologies evolved and their impact on society became clear, ethical issues gained significance.

3. Guiding Principles: The establishment of ethical frameworks, norms, and principles aiming to ensure that AI serves humanity's best interests.

Insights Gained Along the Way

1. Human-Centric Approach: The core principle of ethical AI is to prioritize the well-being and interests of individuals over technological advancements.

2. Transparency and Explainability: Transparent AI systems are vital for developing trust, as consumers need to understand how decisions are made.

3. Bias and justice: The voyage has emphasised the crucial need of tackling biases in AI systems to ensure justice and equitable treatment.

4. Collaboration is Key: Ethical AI development involves collaboration among many stakeholders, including industry, governments, academia, and civil society.

Accomplishments in Ethical AI

1. Ethical Frameworks: The introduction of ethical frameworks and principles has offered a roadmap for responsible AI development.

2. Explainable AI: Significant progress has been made in creating strategies to make AI systems more transparent and explainable.

3. Public Awareness: The voyage has elevated public awareness about AI's ethical consequences, empowering individuals to demand ethical considerations.

Challenges and Unfinished Business

1. Bias Mitigation: Addressing biases remains a complicated task, as AI systems are often taught on skewed data.

2. Regulatory Adaptation: The fast-paced advancement of AI technology offers problems for regulatory frameworks to keep pace with innovations.

3. Global Harmonization: Achieving global agreement in ethical AI legislation and norms remains a problem owing to various cultural and legal contexts.

The Imperative of Continued Dedication

1. ongoing Learning: Ethical AI is an ever-evolving discipline, necessitating ongoing learning, adaptation, and staying informed of developing difficulties.

2. Striking Balance: The journey is about striking a balance between innovation and ethical issues to guarantee that AI actually benefits society.

3. Shared obligation: Ethical AI is a shared obligation among governments, industries, academics, and citizens to create a more just and equitable future.

Ethical AI's Broader Impact

1. Positive Societal Change: Ethical AI contributes to positive societal change by designing technology that enhances human well-being and protects rights.

2. Public Trust: Responsible AI development builds public trust, enabling wider usage and acceptance of AI technology.

3. Influence on Innovation: Ethical considerations promote innovation that corresponds with societal ideals, leading to sustainable and equitable progress.

The Journey Ahead

1. Emerging Technologies: As new technologies emerge, the ethical journey continues with an emphasis on addressing ethical challenges particular to those technologies.

2. Education and Advocacy: Educating the next generation and pushing for ethical AI will be important to perpetuate responsible innovation.

Conclusion

The observations on the voyage of ethical AI indicate a route marked by ongoing development, obstacles, and progress. From identifying the need for ethical considerations to the formation of frameworks and standards, the path shows the dedication to establishing a future where AI technologies serve humanity's best interests. Ethical AI is not a destination but a continual expedition, requiring steadfast commitment to innovation that respects human values, safeguards rights, and supports societal well-being. As the trip unfolds, the lessons acquired and the ideals supported push us toward a more equal, transparent, and responsible technological world.

2.   Looking Ahead: Challenges and Opportunities in the Ethical AI Landscape

The future of artificial intelligence (AI) is both bright and daunting, with tremendous chances to alter businesses and enhance lives. As we peek into the future, it's crucial to grasp the complexities and uncertainties that lie ahead in the ethical AI landscape. This chapter addresses the predicted challenges, the potential chances for positive effect, and the techniques that will be vital in steering the route toward responsible and beneficial AI development.

Anticipated Challenges in the Ethical AI Landscape

1. Bias and Fairness: The challenge of tackling bias in AI systems remains critical, as skewed data can perpetuate unfair outcomes and increase inequities.

2. legal Agility: The rapid progress of AI technology necessitates legal frameworks that can react swiftly to new advancements while maintaining ethical considerations.

3. Privacy Concerns: Striking a balance between AI-driven insights and user privacy remains an issue, as more data is collected for training and decision-making.

4. Transparency and Explainability: Developing AI systems that are transparent, explainable, and understandable to non-experts raises technological and ethical obstacles.

5. Global Harmonization: Achieving a unified set of global AI legislation and standards that respect varied cultural contexts and values is a challenging endeavour.

Opportunities for Positive Impact

1. Healthcare Revolution: Ethical AI can transform healthcare, enhancing diagnoses, treatment, and patient outcomes through tailored medication.

2. Education Enhancement: AI-driven personalized learning systems offer chances to change education by responding to individual needs and learning styles.

3. Environmental Conservation: AI's data analytic capabilities can be utilised to monitor climate change, protect ecosystems, and promote sustainability.

4. Humanitarian Aid: AI-driven solutions can aid in disaster response, resource allocation, and predicting humanitarian emergencies, saving lives and resources.

Strategies for Navigating the Ethical AI Landscape

1. Ethical Literacy: Empowering individuals with a broader awareness of AI ethics guarantees that users can make informed decisions and advocate for responsible AI.

2. Interdisciplinary Collaboration: Effective AI solutions require collaboration amongst experts in technology, ethics, law, social sciences, and beyond.

3. Public Involvement: Including different public perspectives in AI development and policy-making guarantees that technology serves society's best interests.

4. Ethical Technology Leadership: Industry leaders must set ethical standards, encourage responsible practices, and drive positive change across the sector.

Fostering Global Collaboration

1. International Standards: Global collaboration is vital for developing consistent ethical rules and standards that transcend borders.

2. Cross-Cultural Considerations: Recognizing cultural variations and values is crucial in designing ethical AI solutions that resonate with varied populations.

3. Ethics in Emerging Technologies: As technologies like quantum computing and biotechnology evolve, emphasising ethics from the inception is vital.

Ethical AI in Everyday Life

1. Personal gadgets: Ethical AI ensures that AI-powered personal gadgets respect user privacy and give clear insights.

2. Smart Cities: AI-driven urban planning can boost infrastructure, transportation, and resource management while addressing social requirements.

3. Workplace Ethics: AI impacts the workplace through automation and decision-making; ethical issues assure equitable treatment of employees.

The Ongoing Evolution of Ethical AI

1. Continuous Learning: Ethical AI is a growing field; experts must stay knowledgeable about new problems and best practices.

2. Ethical AI Regulation: Governments play a significant role in adapting and enforcing legislation to guide AI development in a responsible path.

Conclusion

The future of ethical AI presents both problems and extraordinary opportunity for beneficial societal impact. Navigating this situation needs a commitment to transparency, fairness, and responsibility. As AI technologies continue to influence our world, it is the collaborative obligation of industry, academia, governments, and society to ensure that these technologies are created and used in ways that line with ethical principles. By grasping the opportunities and addressing the obstacles, we can harness the full potential of AI to create a future that benefits all of humanity, respects individual rights, and maintains ethical principles.

3.   Empowering Readers to Shape the Future of Ethical AI: A Call to Action

In the ever-evolving environment of artificial intelligence (AI), the ability to create the future lies not only in the hands of experts and industry leaders but also in the informed minds of readers and learners. The trip through the world of ethical AI has been explored, the problems and potential disclosed, and the imperative of responsible innovation highlighted. As we complete this tale, it's crucial to remember that every reader has a pivotal role to play in the direction of AI's evolution. This chapter underlines the importance of reader empowerment, gives tangible strategies for promoting ethical AI growth, and reinforces the shared obligation to create a future that represents our highest ethical ambitions.

Harnessing Knowledge for Empowerment

1. Education as Empowerment: Understanding the ethical dimensions of AI empowers readers to make educated decisions, demand transparency, and advocate for responsible actions.

2. Ethical Literacy: Becoming ethically literate in AI offers readers with the critical thinking abilities to handle the complexity of AI's impact on society.

Taking Action as Informed Readers

1. Demand for Transparency: As informed readers, humans may demand transparency in AI systems, ensuring they understand how decisions are made.

2. Advocacy for Fairness: Readers can campaign for fair AI systems that prevent biases, prejudice, and unjust treatment of individuals.

3. Ethical Consideration in Consumption: By supporting ethical AI products and services, readers encourage responsible behaviours throughout the sector.

Collaboration Across Disciplines

1. Interdisciplinary Exchange: Readers from varied backgrounds can collaborate to bring multifaceted ideas to AI ethical talks.

2. Ethics in Tech Development: Encouraging collaboration between tech professionals and ethics specialists guarantees that AI systems are built responsibly.

Guiding Ethical AI Innovation

1. User Input: As users of AI technologies, readers can contribute input on how systems affect their lives, determining the direction of AI research.

2. Participation in Policy Debates: Engaging in policy conversations enables readers to impact legislation and norms that guide ethical AI.

Fostering Ethical AI Mindsets

1. Ethics in AI Curricula: Readers in academia can urge for AI ethics to be integrated into curricula, ensuring the next generation of professionals is ethically competent.

2. Public Awareness Initiatives: Organizing workshops, seminars, and public speeches on AI ethics helps raise awareness and create educated discussions.

Ethical Considerations in Professional Roles

1. Industry Advocates: Professionals across industries can advocate for ethical AI practices within their firms, influencing ethical conduct from within.

2. Tech Innovators: Tech innovators can prioritize ethics, ensuring their creations uphold principles that benefit society and prevent possible harm.

The Broader Impact of Reader Empowerment

1. Cultural Transformation: Reader empowerment encourages a cultural shift that emphasises transparency, responsibility, and ethical behavior in AI development.

2. Policy Influence: Informed readers collectively exert influence over policy formulation, pushing rules that reflect ethical AI objectives.

A Shared Responsibility

1. Collaborative Efforts: Ethical AI's future is a collective undertaking; collaboration across industries is important to developing a technological landscape that respects humanity's values.

2. Ethical Innovation: Ethical AI pioneers have a responsibility to lead by example, demonstrating that innovation and ethics are not mutually contradictory.

Conclusion

As we complete this investigation of ethical AI, it's apparent that every reader has the power to affect the trajectory of AI's progress. Empowered by information, guided by ethical standards, and unified by a common commitment to responsible innovation, readers hold the keys to crafting a future that leverages AI's powers for the greater benefit. Whether as advocates, policy influencers, or professionals, each individual has a particular role to play in building an AI ecosystem that is transparent, fair, and accountable. Through collective action, we can negotiate the challenges, seize the opportunities, and establish a path that empowers both technology and people to prosper together.

Printed in the USA
CPSIA information can be obtained
at www.ICGtesting.com
LVHW020909051223
765519LV00053B/591

9 781088 210680